The ESSENTIALS®

SO-BYU-377

$ 5·95

STATISTICS II

Emil G. Milewski, Ph.D.

This book is a continuation of *"THE ESSEN-TIALS OF STATISTICS I"* and begins with Chapter 8. It covers the usual course outline of Statistics II. Earlier/basic topics are covered in *"THE ESSENTIALS OF STATISTICS I"*.

Research and Education Association
61 Ethel Road West
Piscataway, New Jersey 08854

THE ESSENTIALS® OF STATISTICS II

Revised Printing, 1993

Printed in the United States of America

Library of Congress Catalog Card Number 88-64175

International Standard Book Number 0-87891-659-8

WHAT "THE ESSENTIALS" WILL DO FOR YOU

This book is a review and study guide. It is comprehensive and it is concise.

It helps in preparing for exams, in doing homework, and remains a handy reference source at all times.

It condenses the vast amount of detail characteristic of the subject matter and summarizes the **essentials** of the field.

It will thus save hours of study and preparation time.

The book provides quick access to the important facts, principles, theorems, concepts, and equations of the field.

Materials needed for exams, can be reviewed in summary form-eliminating the need to read and re-read many pages of textbook and class notes. The summaries will even tend to bring detail to mind that had been previously read or noted.

This "ESSENTIALS" book has been prepared by an expert in the field, and has been carefully reviewed to assure accuracy and maximum usefulness.

Dr. Max Fogiel
Program Director

CONTENTS

CHAPTER 8

SAMPLING THEORY

8.1 RANDOM SAMPLING

For a statistician, the relationship between samples and population is important. This branch of statistics is called sampling theory. We gather all pertinent information concerning the sample in order to make statements about the whole population.

Sample quantities such as sample mean, deviation, etc. are called sample statistics or statistics. Based on these quantities we estimate the corresponding quantities for population, which are called population parameters or parameters. For two different samples the difference between sample statistics can be due to chance variation or some significant factor. The latter case should be investigated and possible mistakes corrected. The statistical inference is a study of inferences made concerning a population and based on the samples drawn from it.

Probability theory evaluates the accuracy of such inferences. The most important initial step is the choice of samples which are representative of a population. The methods of sampling are called the design of the experiment. One of the most widely used methods is random sampling.

Random Sampling

A sample of n measurements chosen from a population N ($N > n$) is said to be a random sample if every different sample of the same size n from the population has an equal probability of being selected.

One way of obtaining a random sample is to assign to each member of the population a number. The population becomes a set of numbers. Then, using the random number table, we can choose a sample of desired size.

EXAMPLE:

Suppose 1,000 voters are registered and eligible to vote in an upcoming election. To conduct a poll you need a sample of fifty persons. To each voter you assign a number between one and 1,000. Then, using the random number table or a computer program, you choose at random fifty numbers, which are fifty voters. This is your required sample.

Sampling With and Without Replacement

From a bag containing ten numbers from 1 to 10 we have to draw three numbers. As the first step, we draw a number. Now we have the choice of replacing or not replacing the number into the bag. If we replace the number, then this number can come up again. If the number is not replaced, then it can come up only once.

Sampling where each element of a population may be chosen more than once (i.e. where chosen element is replaced) is called sampling with replacement. Sampling without replacement takes place when each element of a population can be chosen only once.

Remember that populations can be finite or infinite.

EXAMPLE:

A bag contains ten numbers. We choose two numbers without replacement. This is sampling from a finite population.

EXAMPLE:

A coin is tossed ten times and the number of tails is counted. This is sampling from an infinite population.

8.2 SAMPLING DISTRIBUTIONS

A population is given from which we draw samples of size n, with or without replacement. For each sample we compute a statistic such as the mean, standard deviation, variance, etc. These numbers will depend on the sample and they will vary from sample to sample. In this way we obtain a distribution of the statistic which is called sampling distribution.

For example, if for each sample we measure its mean, then the distribution obtained is the sampling distribution of means. In the same way we obtain the sampling distributions of variances, standard deviations, medians, etc.

8.3 THE CENTRAL LIMIT THEOREM

A population is given with a finite mean μ and a standard deviation σ. Random samples of n measurements are drawn. If the population is infinite or if sampling is with replacement, then the relative frequency histogram for the sample means will be approximately normal with mean μ and standard deviation $\dfrac{\sigma}{\sqrt{n}}$.

Suppose the distribution of x for the population is as shown in

Fig. 1, with the mean μ.

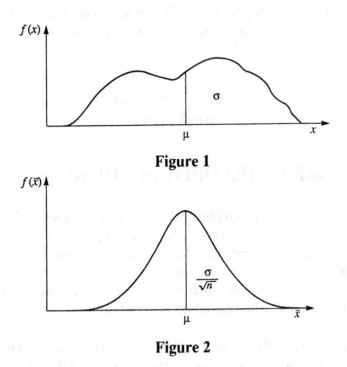

Figure 1

Figure 2

The standard deviation is σ.

Figure 2 shows the relative frequency histogram, called the sampling distribution, for the sample mean $\bar{x}$. The samples with replacement of size n are measured and sample mean $\bar{x}$ is computed. The mean for the sampling distribution of $\bar{x}$ is μ, the same as for the whole population. The standard deviation of the sampling distribution is equal to the standard deviation of the x measurements divided by $\sqrt{n}$, that is $\dfrac{\sigma}{\sqrt{n}}$.

If the samples of size n are drawn without replacement from a finite population of size N, then

$$\mu_n = \mu \quad \text{and} \quad \sigma_n = \frac{\sigma}{\sqrt{n}} \cdot \sqrt{\frac{N-n}{N-1}}$$

where μ and σ denote the population mean and standard deviation; while μ_n and σ_n denote the mean and standard deviation respectively of the sampling distribution.

EXAMPLE:

Suppose a population consists of the five numbers 2, 4, 5, 6, 8. All possible samples of size two are drawn with replacement. Thus, there are $5 \cdot 5 = 25$ samples. The mean of the population is

$$\mu = \frac{25}{5} = 5$$

and the standard deviation of the population is

$$\sigma^2 = \frac{(2-5)^2 + (4-5)^2 + (5-5)^2 + (6-5)^2 + (8-5)^2}{5} = 4$$

$$\sigma = 2$$

We shall list all 25 samples and their corresponding sample means

(2, 2); 2   (4, 2); 3   (5, 2); 3.5 (6, 2); 4   (8, 2); 5
(2, 4); 3   (4, 4); 4   (5, 4); 4.5 (6, 4); 5   (8, 4); 6
(2, 5); 3.5 (4, 5); 4.5 (5, 5); 5   (6, 5); 5.5 (8, 5); 6.5
(2, 6); 4   (4, 6); 5   (5, 6); 5.5 (6, 6); 6   (8, 6); 7
(2, 8); 5   (4, 8); 6   (5, 8); 6.5 (6, 8); 7   (8, 8); 8

The mean of the sampling distribution of means is

$$\mu_2 = \frac{\text{sum of means}}{25} = \frac{125}{25} = 5$$

Thus

$$\mu = \mu_2$$

The variance σ_2^2 is

$$\sigma_2^2 = \frac{\sum (\mu_n - \mu_2)^2}{25} = \frac{50}{25} = 2$$

and $\sigma_2 = 1.414$.

For finite populations and sampling with replacement (or infinite populations) we have

$$\sigma_2 = \frac{\sigma}{\sqrt{n}} = \frac{2}{\sqrt{2}} = 1.414$$

which agrees with the calculated value.

8.4 SAMPLING DISTRIBUTION OF PROPORTIONS

Suppose that for an infinite population the probability of occurrence of an event (i.e. its success) is p. Then the probability of its failure is $q = 1 - p$.

EXAMPLE:

The population is all possible tosses of a coin (we always assume a coin, a die, etc. to be fair). The probability of the event "tails" is $p = \dfrac{1}{2}$.

All possible samples of size n are drawn from this population and for each sample the proportion P of success is determined. For n tosses of the coin, P is the proportion of tails obtained. We have a sampling distribution of proportions. Its mean μ_p and standard deviation σ_p are given by

$$\mu_p = p$$

$$\sigma_P = \sqrt{\frac{p(1-p)}{n}} = \sqrt{\frac{pq}{n}}$$

The above equations are valid for a finite population and sampling with replacement. For large values of n the sampling distribution is close to the normal distribution. For finite populations and sampling without replacement the mean μ_p is

$$\mu_p = p$$

and the standard deviation σ_p is

$$\sigma_P = \sqrt{p(1-p)}$$

8.5 SAMPLING DISTRIBUTIONS OF SUMS AND DIFFERENCES

Two populations are given. From the first population we draw samples of size n_1 and compute a statistic s_1. Thus, we obtain a sampling distribution for the statistic s_1 with the mean μ_{s_1} and standard deviation σ_{s_1}.

Similarly, from the second population we draw samples of size n_2, compute a statistic s_2 and find μ_{s_2} and σ_{s_2}. From all possible combinations of these samples from the two populations we can determine a distribution of the sums, $s_1 + s_2$, which is called

111

the sampling distribution of sums of the statistics. We can also find a distribution of the differences, $s_1 - s_2$, which is called the sampling distribution of differences of the statistics. The mean and standard deviation of the sampling distribution is

for the sum

$$\mu_{s_1+s_2} = \mu_{s_1} + \mu_{s_2}$$

$$\sigma_{s_1+s_2} = \sqrt{\sigma_{s_1}^2 + \sigma_{s_2}^2}$$

for the difference

$$\mu_{s_1-s_2} = \mu_{s_1} - \mu_{s_2}$$

$$\sigma_{s_1-s_2} = \sqrt{\sigma_{s_1}^2 + \sigma_{s_2}^2}$$

The samples have to be independent, that is, do not depend on each other.

EXAMPLE

Suppose f_1 can be any of the elements of the population 3, 5, 7 and f_2 any of the elements of the population 2, 4. Then

$$\mu_{f_1} = \text{mean of population } f_1 = \frac{3+5+7}{3} = 5$$

$$\mu_{f_2} = \text{mean of population } f_2 = \frac{2+4}{2} = 3$$

Now, let us consider the population consisting of the sums of any number of f_1 and any number of f_2.

$$
\begin{array}{ccc}
3+2 & 5+2 & 7+2 \\
3+4 & 5+4 & 7+4
\end{array}
$$

or

$$\begin{array}{ccc} 5 & 7 & 9 \\ 7 & 9 & 11 \end{array}$$

The mean $\mu_{f_1 + f_2}$ is

$$\mu_{f_1 + f_2} = \frac{48}{6} = 8$$

This result is in agreement with the general rule

$$8 = \mu_{f_1 + f_2} = \mu_{f_1} + \mu_{f_2} = 5 + 3$$

The standard deviations are

$$\sigma_{f_1}^2 = \frac{(3-5)^2 + (5-5)^2 + (7-5)^2}{3} = 2.667$$

$$\sigma_{f_1} = 1.633$$

$$\sigma_{f_2}^2 = \frac{(2-3)^2 + (4-3)^2}{2} = 1 \quad \text{and} \quad \sigma_{f_2} = 1$$

Similarly, we compute $\sigma_{f_1 + f_2}$

$$\sigma_{f_1 + f_2}^2 = \frac{(5-8)^2 + (7-8)^2 + (9-8)^2 + (7-8)^2 + (9-8)^2 + (11-8)^2}{6}$$

$$= 3.667$$

$$\sigma_{f_1 + f_2} = 1.915$$

That agrees with the general formula

$$\sigma_{f_1 + f_2} = \sqrt{\sigma_{f_1}^2 + \sigma_{f_2}^2}$$

for independent samples.

Suppose s_1 and s_2 are the sample means from the two populations, which we denote by $\bar{x}_1$ and $\bar{x}_2$. Then, for infinite populations (or finite populations and sampling with replacement) with means μ_1 and μ_2 and standard deviations σ_1 and σ_2 respectively, the sampling distribution of the sums (or the differences) of means is

for sums

$$\mu_{\bar{x}_1 + \bar{x}_2} = \mu_{\bar{x}_1} + \mu_{\bar{x}_2} = \mu_1 + \mu_2$$

$$\sigma_{\bar{x}_1 + \bar{x}_2} = \sqrt{\sigma_{\bar{x}_1}^2 + \sigma_{\bar{x}_2}^2} = \sqrt{\frac{\sigma_1^2}{n_1} + \frac{\sigma_2^2}{n_2}}$$

for differences

$$\mu_{\bar{x}_1 - \bar{x}_2} = \mu_{\bar{x}_1} - \mu_{\bar{x}_2}$$

$$\sigma_{\bar{x}_1 - \bar{x}_2} = \sqrt{\sigma_{\bar{x}_1}^2 + \sigma_{\bar{x}_2}^2} = \sqrt{\frac{\sigma_1^2}{n_1} + \frac{\sigma_2^2}{n_2}}$$

where n_1 and n_2 are sizes of samples.

EXAMPLE:

Two producers manufacture tires. The mean lifetime of tires made by A is 120,000 miles with a standard deviation 20,000 miles, while the mean lifetime of tires made by B is 80,000 miles with a standard deviation 10,000 miles.

Random samples of 200 tires of each brand are tested. What is the probability that tires made by A will have a mean lifetime which is at least 45,000 miles more than the tires made by B? $\bar{x}_A$ and $\bar{x}_B$ denote the mean lifetimes of samples A and B respectively. Then

$$\mu_{\bar{x}_A - \bar{x}_B} = \mu_{\bar{x}_A} - \mu_{\bar{x}_B} = 120,000 - 80,000 = 40,000 \text{ miles}$$

$$\sigma_{\bar{x}_A - \bar{x}_B} = \sqrt{\frac{\sigma_A^2}{n_A} + \frac{\sigma_B^2}{n_B}} = \sqrt{\frac{(20,000)^2}{200} + \frac{(10,000)^2}{200}}$$

$$= 1581$$

The standardized variable for the difference in means is

$$z = \frac{(\bar{x}_A - \bar{x}_B) - (\mu_{\bar{x}_A - \bar{x}_B})}{\sigma_{\bar{x}_A - \bar{x}_B}} = \frac{(\bar{x}_A - \bar{x}_B) - 40,000}{1581}$$

For large samples the distribution is normal. For the difference 45,000 miles

$$z = \frac{45,000 - 40,000}{1581} = 3.16$$

Hence

required probability =
area under normal curve to right of $z = 3.16 =$

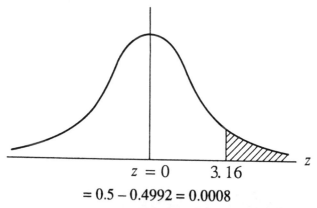

$$= 0.5 - 0.4992 = 0.0008$$

8.6 STANDARD ERRORS

The standard deviation of a sampling distribution of a statistic is called its standard error. If the sample size n is large enough, then the sampling distributions are nearly normal. Statistical methods dealing with such samples are called large sampling methods. When $n < 30$, the samples are small and the theory of small samples, or exact sampling theory, is applied. We shall denote

	Population	Sample
Mean	μ	$\bar{x}$
Standard Deviation	σ	s
Proportion	p	P
rth moment	μ_r	m_r

If the population parameters are unknown we can obtain close estimates from their corresponding sample statistics. Assuming the samples are large enough.

CHAPTER 9

STATISTICAL INFERENCE: LARGE SAMPLES

9.1 ESTIMATION OF PARAMETERS

In most statistical applications we move from a sample to the population. First we draw a sample from its population in accordance with sampling theory. Then, all sample statistics (like sample mean, standard deviation, etc.) are measured which are necessary to estimate population parameters (like population mean, standard deviation, etc.).

Statistical inference deals with the estimation of population parameters from the corresponding sample statistics.

In general all inferences belong to one of the two categories.

EXAMPLE:

The new law is about to be introduced in the State Senate. A questionnaire is given to a sample of residents asking them for possible amendments to this law. Since the law is new no information is available concerning the attitudes in the past. Thus, it is impossible to estimate the average response. But having the sample information we can estimate the response of all residents.

EXAMPLE:

A manufacturer tests a new electric motor for its washing machine. On the average the old one could work 7,000 hours without repair, $\mu = 7,000$. If the new motor is better than the old one it should on the average work longer than 7,000 hours without repairs. Thus, we are testing a hypothesis that for a new motor $\mu > 7,000$. The two examples illustrate the two different inference-making procedures used:

1. estimation

2. test of hypothesis.

Unbiased and Biased Estimates

If for the sampling distribution the mean of a statistic is equal to the corresponding population parameter, then the statistic is called an unbiased estimator of the parameter, otherwise it is called a biased estimator.

EXAMPLE:

The sample mean $\bar{x}$ is an unbiased estimate of the population mean μ because the mean of the sampling distribution of means

$$\mu_{\bar{x}} = \mu$$

EXAMPLE:

The sample variance s^2 is a biased estimate of the population variance σ^2. The mean of the sampling distribution of variance is given by

$$\mu_{s^2} = \frac{n-1}{n} \sigma^2$$

where σ^2 is the population variance and n is the sample size.

9.2 POINT ESTIMATION OF μ

An estimate of a population parameter given by a single number is called a point estimate of the parameter. An estimate of a population parameter given by two numbers between which the parameter lies is called an interval estimate of the parameter.

EXAMPLE:

The distance is given as 3.45 miles. That is a point estimate. If the distance is given as 3.45 ± 0.02 miles then we have an interval estimate. Here the distance and the accuracy of measurement are given. The distance is between 3.43 and 3.47 miles.

In statistical inference-making procedures we not only make an inference but also supply information on how good the inference is. For example, from the measurements of a sample we obtain the value of $\bar{x}$ which is a point estimate of μ. But there are many possible values of x and it may happen that our $\bar{x}$ will not be exactly equal to μ. Thus, we have to supply another piece of information, which is how close $\bar{x}$ is to the parameter we are trying to estimate. The Central Limit Theorem for $\bar{x}$ states that under certain conditions the sampling distribution of $\bar{x}$ would be approximately normal, with mean μ the same as for the population and standard deviation $\dfrac{\sigma}{\sqrt{n}}$, where σ is a standard deviation of a population and n is the size of samples. For the Central Limit Theorem to hold, the sample size must be

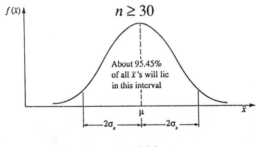

Having the sampling distribution of $\overline{x}$ and areas under the normal curve we know that about 95.45% of the point estimates calculated from repeated samples will be within

$$2\sigma_{\overline{x}} = 2\,\frac{\sigma}{\sqrt{n}}\ \text{of}\ \mu$$

as shown in the figure.

This information tells us how good our point estimate is.

Error of Estimation

For a given point estimate the error of estimation is defined to be

$$\text{error of estimation} = \left|\,\overline{x} - \mu\,\right|$$

Bound on Error

By the Empirical Rule, the error of estimation will be less than $2\sigma_{\overline{x}}$ approximately 95.45% of the time. The quantity

$$2\sigma_{\overline{x}}$$

is called a bound on the error of estimation. The smaller the bound on the error of estimation, the better the inference.

Point Estimation of a Population Mean μ	
Point estimate of μ	$\overline{x}$
Sample size n	$n \geq 30$
Bound on the error	$2\sigma_{\overline{x}} = 2\dfrac{\sigma}{\sqrt{n}}$

EXAMPLE:

A car dealer, who sells new cars is interested in estimating the average number of years a new car can run without any repairs. A random sample of $n = 250$ cars yields the average 2.6 years with a standard deviation of 1.2 years. We have

$$\bar{x} = 2.6 \quad s = 1.2 \quad n = 250$$

Hence our point estimate is $\mu = 2.6$ and the corresponding bound on the error of estimation is given by

$$\text{bound on error} = 2\sigma_{\bar{x}} = \frac{2\sigma}{\sqrt{n}}$$

In this case σ is unknown, but $n \geq 30$ and we can replace σ by s and get an approximate bound on the error

$$2\sigma_{\bar{x}} \approx \frac{2s}{\sqrt{n}} = \frac{2 \cdot 1.2}{\sqrt{250}} = 0.105$$

95% of the sample means calculated in repeated sampling would be within the standard deviations. We can be fairly certain that

$$\bar{x} = 2.6$$

is within 0.1 of the actual mean μ.

9.3 INTERVAL ESTIMATION OF μ

Having a point estimate $\bar{x}$ we can obtain an interval estimate for the population mean μ.

The interval $[\mu - 1.96 \ \sigma_{\bar{x}}, \ \mu + 1.96\sigma_{\bar{x}}]$ includes 95.45% of the x's in repeated sampling, see Fig. 1.

Now consider the interval $[\bar{x} - 1.96\ \sigma_{\bar{x}}, \bar{x} + 1.96\sigma_{\bar{x}}]$. For simplicity, interval $[x - a, x + a]$ will be denoted by $x \pm a$. Whenever $\bar{x} \varepsilon \mu \pm 1.96\ \sigma_{\bar{x}}$ the interval $\bar{x} \pm 1.96\ \sigma_{\bar{x}}$ will contain the parameter μ. This happens 95.45% of times, see Fig. 2.

The interval

$$\bar{x} \pm 1.96\ \sigma_{\bar{x}}$$

represents an interval estimate of μ.

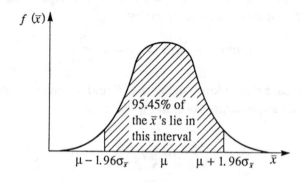

Figure 1

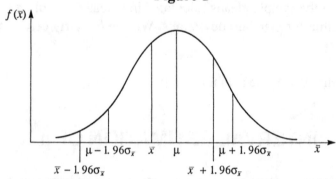

Figure 2

If $\bar{x} \varepsilon \mu \pm 1.96\ \sigma_{\bar{x}}$, then $\mu \varepsilon \bar{x} \pm 1.96\ \sigma_{\bar{x}}$. We measure $\bar{x}$ of a sample and find an interval $\bar{x} \pm 1.96\ \sigma_{\bar{x}}$. The probability that this interval contains μ is 0.95.

Confidence Coefficient

The fraction 0.95 for the interval $\bar{x} \pm 1.96\ \sigma_{\bar{x}}$ is called the confidence coefficient. In repeated sampling, 95% of the time the intervals calculated $\bar{x} \pm 1.96\ \sigma_{\bar{x}}$ will contain the mean μ.

In practice we calculate $\bar{x}$ and only one interval $\bar{x} \pm 1.96\ \sigma_{\bar{x}}$. This interval, called a 95% confidence interval, represents an interval estimate of μ. Of course one can construct many different confidence intervals for μ, depending on the chosen confidence coefficient.

The interval

$$\mu \pm 2.58\ \sigma_{\bar{x}}$$

would include 99% of the values of $\bar{x}$ in repeated sampling.

The interval

$$\bar{x} \pm 2.58\ \sigma_{\bar{x}}$$

forms a 99% confidence interval for μ.

EXAMPLE:

For a random sample of $n = 40$ students, the average weight was 172 lbs. with a standard deviation 54 lbs. Thus

$$\bar{x} = 172 \qquad s = 54 \qquad n = 40$$

The 95% confidence interval is given by

$$\bar{x} \pm 1.96\ \sigma_{\bar{x}}$$

where $\sigma_{\bar{x}} = \dfrac{\sigma}{\sqrt{n}}$.

We shall substitute the sample standard deviation s for σ.

Hence, the 95% confidence interval is

$$172 \pm \frac{1.96 \cdot 54}{\sqrt{40}} = 172 \pm 16.8$$

and the 99% confidence interval is

$$172 \pm \frac{2.58 \cdot 54}{\sqrt{40}} = 172 \pm 22$$

We define

$$\text{confidence coefficient} = 1 - \alpha$$

where $0 \leq \alpha \leq 1$.

For a specific value of $1 - \alpha$, a $100(1 - \alpha)\%$ confidence interval for μ is given by

$$\overline{x} \pm z_{\frac{\alpha}{2}} \sigma_{\overline{x}}$$

The value of $z_{\frac{\alpha}{2}}$ is such that at a distance of $z_{\frac{\alpha}{2}}$ standard deviations to the right of μ, the area under the normal curve is $\frac{\alpha}{2}$.

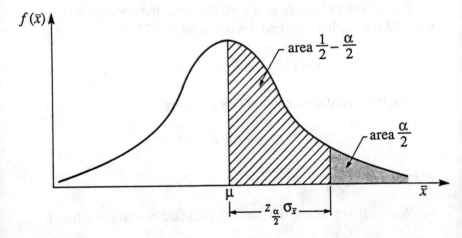

124

The table below gives the confidence coefficient and the corresponding values of $z_{\frac{\alpha}{2}}$

Confidence Coefficient $1 - \alpha$	$z_{\frac{\alpha}{2}}$
0.9973	3
0.99	2.58
0.98	2.33
0.96	2.05
0.95	1.96
0.90	1.645
0.80	1.28
0.683	1
0.5	0.6745

9.4   TEST OF HYPOTHESIS

We shall consider a statistical test of a hypothesis. This test should answer the question, "Is the population mean equal to a specified value μ_0?" A statistical test consists of the following parts:

1. Research hypothesis, denoted by h

2. Null hypothesis, denoted by h_0

3. Test statistic

4. Rejection (or acceptance) region

5. Conclusion

For example we are interested in what profit a one thousand dollar deposit would yield in different banks.

The research hypothesis is that the mean yield per $1,000 is greater than $125, the average observed for banks in the past three years. We formulate the null hypothesis, that

$$\mu = 125$$

Research hypothesis $\qquad\qquad\qquad \mu > 125$

Null hypothesis $\qquad\qquad\qquad\qquad \mu = 125$

To verify the research hypothesis, we must contradict the null hypothesis. A random sample of deposits gives $\bar{x}$ and s, that is the sample mean and standard deviation, respectively.

Whether we accept or reject the null hypothesis is based on a test statistic computed from the sample data. The value $\bar{x}$ can be chosen as the test statistic. Assuming the null hypothesis is true, the sampling distribution of $\bar{x}$ is approximately normal with mean μ. The values $\bar{x}$ located in the right slope of the distribution will be contradictory to the null hypothesis and in favor of the research hypothesis. These values form a rejection region for our statistical test.

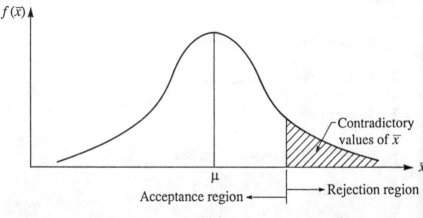

If the observed value of $\bar{x}$ falls in the rejection region, we reject the null hypothesis and accept the research hypothesis.

Two kinds of errors may occur.

1. False rejection of the null hypothesis. The probability of this error is denoted by α.

2. False acceptance of the null hypothesis.

Usually we have to predetermine the value of α. For example, by setting $\alpha = \dfrac{1}{20} = 0.05$ we will be incorrectly rejecting the null hypothesis one time in twenty.

EXAMPLE:

The research hypothesis is that the mean yield per $1,000 is greater than $125. Hence the null hypothesis is $\mu = 125$. A sample of $n = 49$ deposits lead to the average yield $\bar{x} = \$138$ and $s = 37$. Is the research hypothesis true?

We shall set $\alpha = 0.025$

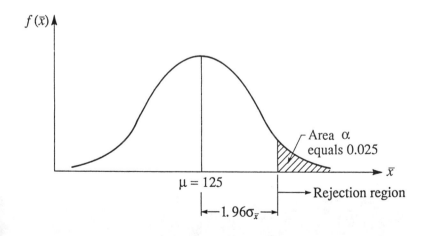

127

$$\sigma_{\bar{x}} \approx \frac{S}{\sqrt{n}} = \frac{37}{\sqrt{49}} = 5.29$$

For $\alpha = 0.025$ we shall reject the null hypothesis if $\bar{x}$ lies more than 1.96 standard deviations above $\mu = 125$.

$$z = \frac{\bar{x} - \mu}{\sigma_{\bar{x}}} = \frac{138 - 125}{5.29} = 2.46$$

We see that the observed value of $\bar{x}$ lies more than 1.96 standard deviations above the mean 125. Hence, we reject the null hypothesis and accept the research hypothesis that the mean yield on a \$1,000 investment is greater than \$125.

Because the rejection region is located in only one tail (slope), the above test is called a one-tailed test. For the research hypothesis $\mu \neq \mu_0$ we formulate a two-tailed test.

CHAPTER 10

STATISTICAL INFERENCE: SMALL SAMPLE RESULTS

10.1 SMALL SAMPLES

In Chapter 9 situations were discussed where a random sample of 30 or more observations was possible. That is not always the case. For example, if you gather data about earthquakes it is rather difficult to obtain a sample of 30 or more. For large samples with $n \geq 30$, the sampling distributions were approximately normal. For small samples with $n < 30$, this approximation is inaccurate and becomes worse as the number of measurements n decreases.

Small Sampling Theory

A study of sampling distributions of statistics for small samples is called small sampling theory. The results hold for small and large samples. Two important distributions used are:

1. Student's t Distribution

2. Chi-Square Distribution.

10.2 STUDENT'S *t* DISTRIBUTION

For small sample sizes the test statistic

$$z = \frac{\overline{x} - \mu_0}{\sigma} \sqrt{n} \qquad (1)$$

with σ replaced by s is falsely rejecting the null hypothesis $\mu = \mu_0$ at a higher percentage than that specified by α.

W.S. Gosset found the distribution and percentage points of the test statistic

$$\frac{\overline{x} - \mu_0}{s} \sqrt{n}, \quad n < 30$$

He published the results under the pen name Student.

Define the statistic

$$t = \frac{\overline{x} - \mu}{s} \sqrt{n - 1} \qquad (2)$$

Now consider samples of size n drawn from a normal population with mean μ. For each sample compute t using the sample mean $\overline{x}$ and sample standard deviation s. Thus we obtain the sampling distribution for t, which is given by

$$Y = \frac{Y_0}{\left(1 + \dfrac{t^2}{n-1}\right)^{\frac{n}{2}}} = \frac{Y_0}{\left(1 + \dfrac{t^2}{\nu}\right)^{\frac{\nu+1}{2}}} \qquad (3)$$

where Y_0 is a constant depending on the sample size n and such that the area under the curve (3) is equal to one.

The constant

$$\nu = n - 1 \qquad (4)$$

(sometimes denoted df) is called the number of degrees of freedom.

The distribution (3) is called Student's t distribution.

Note that for large values of n curve (3) approaches the standardized normal curve

$$Y = \frac{1}{\sqrt{2\pi}} e^{-\frac{t^2}{2}} \qquad (5)$$

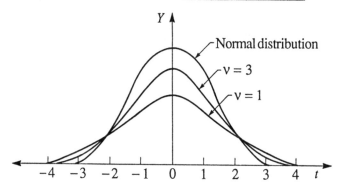

There are many different t distributions. We choose a particular one by specifiying the parameter ν (the number of degrees of freedom).

As n, and hence ν, gets larger, the t distributions approaches the z distribution.

Confidence Intervals

Using the tables of the t distribution we can define 90%, 95%, 99% and other confidence intervals as was done for the normal distributions. The population mean μ can be estimated within specified limits of confidence.

Let us denote by $-t_{0.95}$ and $t_{0.95}$ the values of t for which 5% of the area lies in each tail of the t distribution. Then a 90% confidence interval for t is

$$-t_{0.95} < \frac{\bar{x} - \mu}{s} \sqrt{n-1} < t_{0.95} \qquad (6)$$

From (6) we find

$$\bar{x} - t_{0.95} \frac{s}{\sqrt{n-1}} < \mu < \bar{x} + t_{.95} \frac{s}{\sqrt{n-1}} \qquad (7)$$

Hence, μ lies in the interval $\bar{x} \pm t_{0.95} \dfrac{s}{\sqrt{n-1}}$ with 90% confidence.

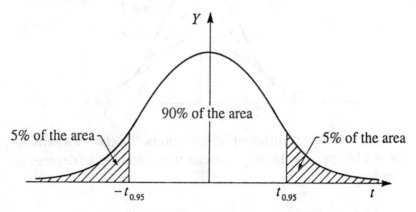

In general, the confidence limits for population means are

defined by

$$\overline{x} \pm t_c \frac{s}{\sqrt{n-1}} \qquad (8)$$

where the values $\pm t_c$ are called critical values or confidence coefficients.

We choose t_c depending on the confidence desired and the sample size.

EXAMPLE:

The graph of Student's t distribution with 7 degrees of freedom ($\nu = 7$) is shown

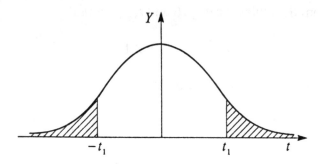

We want to find the value of t_1 for which the total shaded area is 0.05. Student's t distribution is symmetric. Thus, the shaded area to the right of t_1 is 0.025. The area to the left is

$$1 - 0.025 = 0.975$$

t_1 represents the 97.5th percentile $t_{0.975}$.

From the tables for $\nu = 7$ we find

$$t_{0.975} = 2.36$$

The required value of t is 2.36.

A sample of 17 measurements of the diameter of a drill gave a mean $\bar{x} = 3.47$ inches and a standard deviation $s = 0.05$ inches.

We shall find 95% confidence limits for the actual diameter of the drill. The 95% confidence limits are given by

$$\bar{x} \pm t_{0.975} \frac{s}{\sqrt{n-1}}$$

Since $n = 17$, the number of degrees of freedom ν is

$$\nu = n - 1 = 16$$

From the tables for $\nu = 16$ and $t_{0.975}$ we find

$$t_{0.975} = 2.12$$

Substituting all values we find

$$\bar{x} \pm t_{0.975} \frac{s}{\sqrt{n-1}} = 3.47 \pm 2.12 \frac{0.05}{4}$$

$$= 3.47 \pm 0.0265$$

Hence, we are 95% confident that the actual mean lies between 3.4965 inches and 3.4435 inches.

10.3 TESTS OF HYPOTHESES AND SIGNIFICANCE

The tests of hypotheses and significances for small samples are similar to those for large samples. The z score of z statistic is replaced for small samples with the appropriate t score or t statistic.

Small-sample test for μ

Research hypothesis h:

$$1. \quad \mu > \mu_0$$
$$2. \quad \mu < \mu_0$$
$$3. \quad \mu \neq \mu_0$$

Null hypothesis h_0 :

$$\mu = \mu_0$$

Test statistic

$$t = \frac{\bar{x} - \mu}{s} \sqrt{n - 1}$$

where $\bar{x}$ is the mean of the sample of size n.

Rejection region: For a probability α of a type 1 error and $\nu = n - 1$

1. reject h_0 if $t > t_\alpha$,

2. reject h_0 if $t < -t_\alpha$,

3. reject h_0 if $|t| > t_{\frac{\alpha}{2}}$

EXAMPLE:

A manufacturer claims that a mean lifetime of his tires is 75,000 miles. A test was performed on 5 sets of tires, their mean lifetime was 73,500 miles and a standard deviation $s = 1,250$ miles. We want to verify the manufacturer's claim at a level of significance of 0.05.

The null hypothesis h_0 is $\mu_0 = 75,000$. We have

$$t = \frac{\overline{x} - \mu}{s} \sqrt{n - 1} = \frac{73,500 - 75,000}{1,250} \sqrt{5 - 1}$$

$$= -4.8$$

For a one-tailed test at a 0.05 level of significance we shall decide according to the rules:

1. Accept h_0 if t is greater than $-t_{0.95}$. For $v = 5 - 1 = 4$ the value of $-t_{0.95}$ is -2.13.

2. Reject h_0 otherwise. Since $t = -4.8$, we reject h_0.

Differences of Means

Suppose two random samples of sizes n_1 and n_2 are drawn from normal populations with equal standard deviations $\sigma_1 = \sigma_2$. The samples have means x_1, x_2 and standard deviations s_1, s_2 respectively. We formulate the null hypothesis h_0 as follows:

The samples come from the same population, that is

$$\mu_1 = \mu_2 \text{ and } \sigma_1 = \sigma_2.$$

The t score used is given by

$$t = \frac{\overline{x}_1 - \overline{x}_2}{\sigma\sqrt{\dfrac{1}{n_1} + \dfrac{1}{n_2}}}$$

where

$$\sigma = \sqrt{\frac{n_1 s_1^2 + n_2 s_2^2}{n_1 + n_2 - 2}}$$

The distribution is Student's t distribution with

$$\nu = n_1 + n_2 - 2$$

degrees of freedom.

EXAMPLE:

The intelligence quotients of two groups of students from two universities were tested. The test should determine if there is a significant difference between the I.Q.'s of the two groups.

The following results were obtained:

For Group A

sample size $n_A = 14$, mean 146 with a standard deviation of 12

For Group B

sample size $n_B = 17$, mean 141 with a standard deviation of 10

Let μ_A and μ_B denote population mean I.Q.'s of students from two universities.

The level of significance we set at 0.05. We have to decide between the two hypotheses:

1. h_0: $\mu_A = \mu_B$, there is no significant difference between the groups.

2. h_1: $\mu_A \neq \mu_B$, there is a significant difference between the groups.

We have

$$\sigma = \sqrt{\frac{n_A s_A^2 + n_B s_B^2}{n_A + n_B - 2}} = \sqrt{\frac{14 \cdot (12)^2 + 17 \cdot (10)^2}{14 + 17 - 2}}$$

$$= 11.32$$

$$t = \frac{\overline{x}_A - \overline{x}_B}{\sigma \sqrt{\frac{1}{n_A} + \frac{1}{n_B}}} = \frac{146 - 141}{11.32 \sqrt{\frac{1}{14} + \frac{1}{17}}}$$

$$= 1.22$$

For the two-tailed test at a level of significance, we would reject h_0 if t were outside the range $-t_{0.975}$ to $t_{0.975}$. For $\nu = 29$ this range is -2.04 to 2.04. Hence we cannot reject h_0 at a level of significance 0.05.

The conclusion is that there is no difference between the I.Q.'s of the two groups.

10.4 THE CHI-SQUARE DISTRIBUTION

We define the statistic

$$\chi^2 = \frac{(x_1 - \bar{x})^2 + (x_2 - \bar{x})^2 + \ldots + (x_n - \bar{x})^2}{\sigma^2}$$

χ is the Greek letter chi.

Samples of size n are drawn from a normal population with standard deviation σ. For each sample, the value χ^2 is computed. We obtain the chi-square distribution

$$Y = Y_0 \, \chi^{\nu-2} \, e^{-\frac{1}{2}\chi^2}$$

where $\nu = n - 1$ is the number of degrees of freedom, and Y_0 is a constant depending on ν and such that the area under the curve is one.

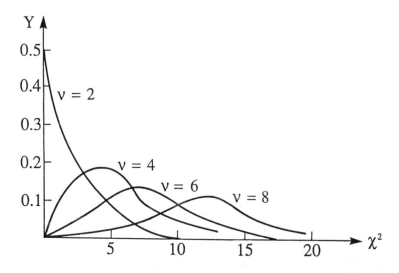

10.5 CONTROL CHARTS

From the manufacturer's and consumer's point of view it is important to maintain a steady level of quality of a product.

Control charts are used to monitor the quality of a product and to detect possible shifts in quality. For samples collected over a period of time we graph the sample mean or sample range.

A control chart consists of three lines: a center line, an upper line and a lower line. The means of successive samples are plotted on the chart.

The center line denoted by $\overline{x}_c$ represents the average of k sample means each computed from m observations. Generally we take

$$k \geq 25$$
$$n \geq 4$$

The samples are taken at the time when production is judged to be normal.

By x_{ij} we denote the jth observation in sample i.

$$\overline{x}_c = \frac{\sum_i \sum_j x_{ij}}{km}$$

The upper control line (UCL) is computed from

$$UCL = \overline{x}_c + 3\frac{\sigma}{\sqrt{n}}$$

and the lower control line (LCL)

$$LCL = \overline{x}_c - 3\frac{\sigma}{\sqrt{n}}$$

The interval $\bar{x}_c \pm 3\,\dfrac{\sigma}{\sqrt{n}}$ should contain almost all the sample means $\displaystyle\sum_j x_{ij}$ in repeated sampling.

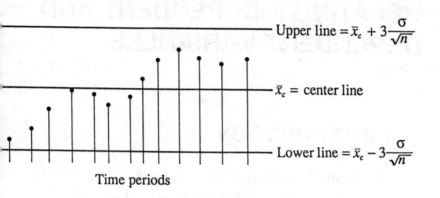

Upper line $= \bar{x}_c + 3\dfrac{\sigma}{\sqrt{n}}$

$\bar{x}_c$ = center line

Lower line $= \bar{x}_c - 3\dfrac{\sigma}{\sqrt{n}}$

Time periods

CHAPTER 11

MATHEMATICAL MODELS RELATING INDEPENDENT AND DEPENDENT VARIABLES

11.1 INTRODUCTION

Often in practice a relationship exists between two or more variables. The dependent variable (the variable of interest) can be affected by one or more independent variables. We usually try to express this relationship in the mathematical form of an equation. The simplest one is the linear equation (equation of straight line).

Straight Line

$$y = a_0 + a_1 x$$

The response y is related to a single quantitative independent variable x. Here a_0 and a_1 are constants. This model is a deterministic model; the value of y is uniquely determined by the value of x.

Parabola

$$y = a_0 + a_1 x + a_2 x^2$$

Cubic Curve

$$y = a_0 + a_1 x + a_2 x^2 + a_3 x^3$$

nth Degree Curve

$$y = a_0 + a_1 x + \ldots + a_n x^n$$

In all those cases a response y is related to a single independent variable. Note that the right hand side of all equations are polynomials.

There are many other possible relationships between x and y.

Hyperbola $\qquad\qquad\qquad y = \dfrac{1}{a_0 + a_1 x}$

Exponential curve $\qquad\quad y = ab^x$

Geometric curve $\qquad\quad y = ax^b$

Logistic curve $\qquad\qquad y = \dfrac{1}{ab^x + d}$

In practice it often happens that the relation between x and y cannot be expressed in the form of an equation. The data can be presented in a scatter diagram.

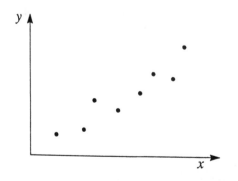

8 points are shown in the diagram, their distribution resembles a straight line. We introduce the following model

$$y = a_0 + a_1 x + \varepsilon$$

where ε is a random error, here it represents the difference between a measurement y and a point on the line $a_0 + a_1 x$. This model is called the probabilistic model.

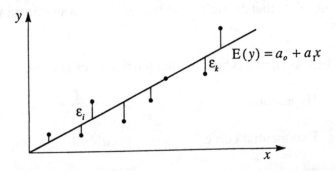

The average of y, called the expected value of y for a fixed x is

$$E(y) = a_0 + a_1 x$$

If two points

$$(x_1, y_1) \text{ and } (x_2, y_2)$$

on the line are given, then the constants in $a_0 + a_1 x = y$ can be determined. The equation of the line can be written

$$y - y_1 = \frac{(y_2 - y_1)}{x_2 - x_1} (x - x_1)$$

11.2 PROBABILISTIC MODELS

The simplest type of probabilistic model is

$$y = a_0 + a_1 x + \varepsilon$$

The assumption is that the average value of ε for a given value of x is

$$E(\varepsilon) = 0$$

Hence, the expected value of y for a given value of x is

$$E(y) = a_0 + a_1 x$$

Not all data would fit this model. There are situations which require

$$y = a_0 + a_1 x + a_2 x^2 + \varepsilon$$

with

$$E(y) = a_0 + a_1 x + a_2 x^2$$

A general polynomial probabilistic model relating a single independent variable x to a dependent variable y is given by

$$y = a_0 + a_1 x + a_2 x^2 + \ldots + a_n x^n + \varepsilon$$

with

$$E(y) = a_0 + a_1 x + \ldots + a_n x^n$$

The choice of n depends on the experimental situation.

Models with Several Independent Variables

Sometimes y depends on more than one independent variable. Here are some typical probabilistic models relating a response y to two independent variables x_1 and x_2:

$$y = a_0 + a_1 x_1 + a_2 x_2 + \varepsilon$$

$$y = a_0 + a_1 x_1 + a_2 x_2 + a_3 x_1 x_2 + \varepsilon$$

$$y = a_0 + a_1 x_1 + a_2 x_1^2 + a_3 x_2 + \varepsilon$$

Similarly we can construct models relating y to three or more independent variables.

The general linear model

$$y = a_0 + a_1 x_1 + a_2 x_2 + \ldots + a_n x_n + \varepsilon$$

where $x_1, \ldots, x_n$ are independent variables, $a_0, a_1, \ldots, a_n$ are unknown parameters, and ε is a random error with

$$E(\varepsilon) = 0$$

All probabilistic models discussed here are special cases of the general linear model. For example the model

$$y = a_0 + a_1 x + a_2 x^2 + a_3 x^3 + \varepsilon$$

is equivalent to a linear model

$$y = a_0 + a_1 x_1 + a_2 x_2 + a_3 x_3 + \varepsilon$$

where $x = x_1$, $x^2 = x_2$, $x^3 = x_3$.

Individual terms in the general model are classified by their exponents. The degree of a term is equal to the sum of the exponents for the variables in this term. Thus

$$6x_4^7 \text{ is a 7th-degree term}$$

and

$$x_3^4 x_4^5 \text{ is a 9th-degree term.}$$

A first-order model is a general linear model that contains all possible first-degree terms in the independent variables.

11.3 THE METHOD OF LEAST SQUARES

Let $x_1, x_2, \ldots, x_n$ be independent variables and y a response

$$y = a_0 + a_1 x_1 + a_2 x_2 + \ldots + a_n x_n + \varepsilon$$

We assume that the random error has expectation zero and obtain the expected value of y

$$E(y) = a_0 + a_1 x_1 + \ldots + a_n x_n$$

This line is called the regression of y on $x_1, \ldots, x_n$. In real situations the parameters $a_0, a_1, \ldots, a_n$ are not known. We cannot find $E(y)$ but we can construct an estimate of $E(y)$ by using the equation

$$\hat{y} = \hat{a}_0 + \hat{a}_1 x_1 + \ldots + \hat{a}_n x_n$$

where $\hat{a}_0, \hat{a}_1, \ldots, \hat{a}_n$ are estimates of the unknown parameters $a_0, a_1, \ldots, a_n$. We find these estimates from sample data.

Consider the probabilistic model

$$y = a_0 + a_1 x + \varepsilon$$

for the linear regression

$$E(y) = a_0 + a_1 x$$

There are many ways of finding an estimate of $E(y)$

$$\hat{y} = \hat{a}_0 + \hat{a}_1 x$$

One can draw an approximating curve to fit a set of data. This method is called a freehand method of curve fitting.

The more reliable method is the method of least squares.

Let $\hat{y}$ denote the predicted value of y for a given value of x.

We define

$$\text{residual} = y - \hat{y}$$

The method of least squares finds the prediction line

$$\hat{y} = \hat{a}_0 + \hat{a}_1 x$$

that minimizes the value of

$$\sum (y - \hat{y})^2$$

The sum is taken over all sample points. For the linear model

$$y = a_0 + a_1 x + \varepsilon$$

this sum is equal to

$$\sum (y - \hat{y})^2 = \sum (y - \hat{a}_0 - \hat{a}_1 x)^2$$

The method of least squares is based on finding the estimates of $\hat{a}_0$ and $\hat{a}_1$ which minimize

$$\sum (y - \hat{y})^2$$

The results are summarized below.

Least Squares Estimates of a_0 and a_1

$$\hat{a}_1 = \frac{\sum (x - \overline{x})(y - \overline{y})}{\sum (x - \overline{x})^2}$$

$$\hat{a}_0 = \overline{y} - \hat{a}_1 \overline{x}$$

Note that

$$\sum (x - \overline{x})(y - \overline{y}) = \sum xy - \frac{(\sum x)(\sum y)}{n}$$

$$\sum (x - \overline{x}) = \sum x^2 - \frac{(\sum x)^2}{n}$$

EXAMPLE:

In some statistical experiments a random sample of size $n = 8$ was chosen and the corresponding values of x and y were measured.

x	y	x^2	xy
37	31	1369	1147
25	19	625	475
32	26	1024	832
18	12	324	216

Table Continued

x	y	x^2	xy
29	27	841	783
34	28	1156	952
23	19	529	437
28	21	784	588
36	30	1296	1080

$\sum x = 262 \quad \sum y = 213 \quad \sum x^2 = 7948 \quad \sum xy = 6510$

We shall construct a straight line which approximates the data of the table. We are looking for the equation of the line

$$y = a_0 + a_1 x$$

We have

$$a_1 = \frac{\sum (x - \bar{x})(y - \bar{y})}{\sum (x - \bar{x})^2} = \frac{n \sum xy - (\sum x)(\sum y)}{n \sum x^2 - (\sum x)^2}$$

$$= \frac{9 \cdot (6510) - (262) \cdot (213)}{9 \cdot (7948) - (262)^2}$$

$$= \frac{2784}{2888} = 0.964$$

$$a_0 = \bar{y} - a_1 \bar{x} = \frac{(\sum x^2)(\sum y^2) - (\sum x)(\sum xy)}{n \sum x^2 - (\sum x)^2}$$

$$= \frac{(7948) \cdot (213) - (262) \cdot (6510)}{9 \cdot (7948) - (262) \cdot (262)}$$

$$= \frac{-12696}{2888} = -4.396$$

Thus

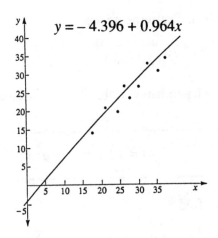

$$y = -4.396 + 0.964x$$

Plot of the least squares equation

Consider again the equation

$$y = a_0 + a_1 x$$

from which we can obtain the normal equations

$$\left.\begin{array}{l} \Sigma y = na_o + a_1 \, \Sigma x \\ \Sigma xy = a_0 \Sigma x + a_1 \, \Sigma x^2 \end{array}\right\}$$

Taking the data from table we find

$$\left.\begin{array}{l} 213 = 9a_0 \quad + 262 \, a_1 \\ 6510 = 262 \, a_0 + 7948 \, a_1 \end{array}\right\}$$

Solving this system simultaneously we find

$$\begin{array}{l} a_0 = -4.396 \\ a_1 = \quad 0.964 \end{array}$$

The Least Square Parabola

Suppose the measurements are

$$(x_1, y_1)\ (x_2, y_2)\ \ldots\ (x_n, y_n)$$

where x is an independent variable. The least square parabola has the equation

$$E(y) = a_0 + a_1 x + a_2 x^2$$

from which we find

$$\sum y = a_0 n + a_1 \sum x + a_2 \sum x^2$$
$$\sum xy = a_0 \sum x + a_1 \sum x^2 + a_2 \sum x^3$$
$$\sum x^2 y = a_0 \sum x^2 + a_1 \sum x^3 + a_2 \sum x^4$$

This system is called the system of the normal equations for the least square parabola.

Solving simultaneously the normal equations we find a_0, a_1, a_2.

General Linear Models

We can apply the method of least squares to estimate parameters of the general linear model. Let

$$y = a_0 + a_1 x_1 + a_2 x_2 + \ldots + a_k x_k + \varepsilon$$

We look for estimates $\hat{a}_0, \hat{a}_1, \ldots, \hat{a}_k$ of $a_0, a_1, \ldots a_k$ respectively, which minimize the value of

$$\sum (y - \hat{y})^2 = \sum (y - \hat{a}_0 - \hat{a}_1 x_1 - \hat{a}_2 x_2 - \ldots - \hat{a}_k x_k)^2$$

152

It can be shown that

$$\sum(y - \bar{y})^2 = \sum(y - \hat{y})^2 + \sum(\hat{y} - \bar{y})^2$$

11.4 LINEAR MODELS IN MATRIX NOTATION

Consider a linear model which relates a response y to a set of k independent variables $x_1, x_2, \ldots x_k$

$$y = a_0 + a_1 x_1 + a_2 x_2 + \ldots + a_k x_k + \varepsilon$$

A sample of n measurements was taken (where $n > k$). For each measurement we obtain a pair (x_i, y_i) $i = 1, \ldots, n$. Hence each individual observation can be written as

$$y_i = a_0 + a_1 x_{i1} + a_2 x_{i2} + \ldots + a_k x_{ik} + \varepsilon_i$$
$$\text{for } i = 1, 2, \ldots, n.$$

ε_i is the random error for the ith response.

Using the matrix notation we can write the set of observations $y_1, y_2, \ldots, y_n$ as a $n \times 1$ matrix Y

$$Y = \begin{bmatrix} y_1 \\ y_2 \\ \vdots \\ y_n \end{bmatrix}$$

and $(x_{11}, x_{12}, \ldots, x_{1k}, x_{21}, \ldots, x_{2k}, \ldots, x_{n1}, \ldots, x_{nk})$ as the $n \times k$ matrix. For reasons, which become clear later, we add one column to this matrix consisting of 1's.

We have the

$$n \times (k + 1)$$

matrix denoted by X

$$X = \begin{bmatrix} 1 & x_{11} & x_{12} & \cdots & x_{1k} \\ 1 & x_{21} & x_{22} & \cdots & x_{2k} \\ \vdots & \vdots & \vdots & & \vdots \\ 1 & x_{n1} & x_{n2} & \cdots & x_{nk} \end{bmatrix}$$

Note that row ith of X contains 1 and the settings of the k independent variables for the ith observation. We define

$$A = \begin{bmatrix} a_0 \\ a_1 \\ \vdots \\ a_k \end{bmatrix}$$

A is a $(k + 1) \times 1$ matrix consisting of the unknown parameters a_0, ... a_k.

Finally the matrix of errors is defined as

$$\varepsilon = \begin{bmatrix} \varepsilon_1 \\ \varepsilon_2 \\ \vdots \\ \varepsilon_n \end{bmatrix}$$

ε is an $n \times 1$ matrix.

In matrix notation the system of equations for the general linear model can be written as

$$
\begin{bmatrix} y_1 \\ y_2 \\ \vdots \\ y_n \end{bmatrix} = \begin{bmatrix} 1 & x_{11} & x_{12} & \cdots & x_{1k} \\ 1 & x_{21} & x_{22} & \cdots & x_{2k} \\ \vdots & \vdots & \vdots & & \vdots \\ 1 & x_{n1} & x_{n2} & \cdots & x_{nk} \end{bmatrix} \begin{bmatrix} a_0 \\ a_1 \\ \vdots \\ a_k \end{bmatrix} + \begin{bmatrix} \varepsilon_1 \\ \varepsilon_2 \\ \vdots \\ \varepsilon_n \end{bmatrix}
$$

or as

$$Y = XA + \varepsilon$$

Multiplying the matrices we multiply rows by columns. Hence

$$y_1 = (\text{first row of X}) \cdot (\text{column of } A) + \varepsilon_1$$

that is

$$y_1 = a_0 + a_1 x_{11} + a_2 x_{12} + \ldots + a_k x_{1k} + \varepsilon_1$$

Similarly

$$y_1 = (i\text{th row of X}) \cdot (\text{column of } A) + \varepsilon_i$$

By $\hat{A}$ we denote the matrix of least squares estimates for the parameters of the general linear model

$$
\hat{A} = \begin{bmatrix} \hat{a}_0 \\ \hat{a}_1 \\ \vdots \\ \hat{a}_k \end{bmatrix}
$$

We define

The Transpose of a Matrix B

The transpose of a matrix B is a new matrix, denoted by B^T obtained from B by interchanging the corresponding rows and columns of the matrix B.

The first row of A becomes the first column of A^T, the second row of A becomes the second column of A^T, and so on.

EXAMPLE:

If

$$A = \begin{bmatrix} 2 & 3 & 5 \\ 1 & 7 & 0 \end{bmatrix}$$

then

$$A^T = \begin{bmatrix} 2 & 1 \\ 3 & 7 \\ 5 & 0 \end{bmatrix}$$

If the dimensions of A are $n \times k$, then the dimensions of A^T are $k \times n$.

The Inverse Matrix

The inverse of a square matrix B, denoted by B^{-1}, is a matrix such that

$$B B^{-1} = I$$
$$B^{-1} B = I$$

when I is the identity matrix

$$I = \begin{bmatrix} 1 & 0 & \cdot & \cdot & \cdot & 0 \\ 0 & 1 & & & \cdot & \cdot \\ 0 & & \cdot & & & \cdot \\ \cdot & & & \cdot & & 0 \\ \cdot & \cdot & \cdot & & 1 & 0 \\ 0 & \cdot & \cdot & \cdot & 0 & 1 \end{bmatrix}$$

The diagonal of I consists of 1's and all other elements are zero.

EXAMPLE:

Let

$$B = \begin{bmatrix} 2 & 1 & 3 \\ 1 & 1 & 2 \\ 2 & 1 & 1 \end{bmatrix}$$

We shall find B^{-1}

1. The determinant of B is

$$|B| = -2$$

Hence, since B is a square matrix and the determinant of B is non-zero, B^{-1} exists.

2. The cofactor matrix of B is

$$\begin{bmatrix} -1 & 3 & -1 \\ 2 & -4 & 0 \\ -1 & -1 & 1 \end{bmatrix}$$

Note that a cofactor matrix of B is a square matrix obtained from B by replacing each element by its cofactor.

The cofactor of (i, j) element of a square matrix B is defined as $(-1)^{i+j}$ times the determinant of the matrix formed by removing all the elements of the ith row and the jth column of matrix B.

3. The transpose of the cofactor matrix is

$$\begin{bmatrix} -1 & 2 & -1 \\ 3 & -4 & -1 \\ -1 & 0 & 1 \end{bmatrix}$$

4. Each element of the transpose of the cofactor is divided by the determinant of B

$$|B| = -2$$

The obtained matrix is the inverse of B

$$B^{-1} = \begin{bmatrix} \frac{1}{2} & -1 & \frac{1}{2} \\ -\frac{3}{2} & 2 & \frac{1}{2} \\ \frac{1}{2} & 0 & -\frac{1}{2} \end{bmatrix}$$

Note that

$$BB^{-1} = \begin{bmatrix} 2 & 1 & 3 \\ 1 & 1 & 2 \\ 2 & 1 & 1 \end{bmatrix} \begin{bmatrix} \frac{1}{2} & -1 & \frac{1}{2} \\ -\frac{3}{2} & 2 & \frac{1}{2} \\ \frac{1}{2} & 0 & -\frac{1}{2} \end{bmatrix} = \begin{bmatrix} 1 & 0 & 0 \\ 0 & 1 & 0 \\ 0 & 0 & 1 \end{bmatrix}$$

$$= I$$

and $B^{-1}B = I$.

In this manner we find the inverse of any square matrix which has a non-zero determinant.

Having all necessary definitions we can write the solution of

$$Y = XA + \varepsilon$$

where $\hat{A}$ is the matrix of least squares estimates for the parameters as

$$\hat{A} = (X^T X)^{-1} x^T Y$$

provided $X^T X$ has an inverse.

The solution $\hat{A}$ is the set of parameters

$$\hat{a}_0, \hat{a}_1, ..., \hat{a}_k$$

in the general linear model that minimizes

$$\sum_{i=1}^{n} (y_i - \hat{y}_i)^2$$

Usually the computer programs are used to find the parameters of the linear models with many independent variables.

Regression

Consider the linear model with one independent variable

$$y = a_0 + a_1 x + \varepsilon$$

On the basis of sample data we find the estimates of parameters a_0, a_1 using the least square method. Thus, we can estimate the value of y from a least square curve which fits the sample data. The resulting curve is called a regression curve of y on x (here we estimate y from x).

Similarly we can estimate x from y and use a regression curve of x on y (we simply interchange the variables).

Note that in general the regression curve of x on y and the regression curve of y on x are different.

CHAPTER 12

GENERAL LINEAR MODEL, INFERENCES. CORRELATION THEORY

12.1 INTRODUCTION

A set consisting of experimental data is given. The problem (situation) is described by the general linear model with a prescribed number of independent variables. Using the data we find least square estimates of parameters in this model. The next step is to make inferences about the corresponding population parameters.

12.2 SINGLE PARAMETER

Consider the general linear model with k independent variables and n experimental data

$$y_i = a_0 + a_1 x_{i1} + \ldots + a_k x_{ik} + \varepsilon_i$$

$$i = 1, 2, \ldots, n$$

We assume that

1. The random error ε_i of observation i has expectation zero.

161

2. Random errors $\varepsilon_1, \ldots \varepsilon_n$ are independent of each other.

3. $\varepsilon = [\varepsilon_1, \ldots, \varepsilon_n]$ has mean zero and variance σ^2.

Using these assumptions we find that the distribution of the least squares estimate

$$\hat{a}_j$$

has mean a_j and variance

$$P_{jj}\sigma^2$$

where $j = 1, \ldots, k$ and the constant P_{ij} is the jth diagonal element of

$$(X^T X)^{-1}$$

$$X^T X^{-1} = \begin{bmatrix} P_{00} & & & & \\ & P_{11} & & & \\ & & P_{22} & & \\ & & & \ddots & \\ & & & & P_{kk} \end{bmatrix}$$

The variance of $\hat{a}_0$ denoted by $V\left(\hat{a}_0\right)$ is

$$V\left(\hat{a}_0\right) = P_{00}\sigma^2$$

and the variance of $\hat{a}_j$ is

$$V\left(\hat{a}_j\right) = P_{jj}\sigma^2$$

From the sample data we obtain an estimate of σ^2.

For the general linear model

$$y = a_0 + a_1 x_1 + \ldots + a_k x_k + \varepsilon_i \, , \, E(\varepsilon) = 0$$

we have

$$\varepsilon = y - E(y)$$

Since $E(y)$ is not known, we can estimate $E(y)$ from $\hat{y}$.

The residual sum of squares (also called the sum of squares for error) is defined as

$$\sum (y - \hat{y})^2$$

Dividing $\sum (y - \hat{y})^2$ by the number of degrees of freedom we find an estimate of σ^2.

The number of degrees of freedom is equal to the sample size n minus the number of parameters $k + 1$. Hence

$$\text{number of degrees of freedom} = n - (k + 1).$$

We have an estimate of σ^2 from the general linear model

$$s^2 = \frac{\sum (y - \hat{y})^2}{n - (k + 1))}$$

or using the matrix notation

$$s^2 = \frac{Y^T Y - \hat{A}^T X^T Y}{n - k - 1}$$

Usually we assume that ε_i's in the general linear model are normally distributed. Hence, for a specific parameter we can determine

$$100\,(1-\alpha)\%$$

confidence interval

$$\hat{a}_i \pm t_{\frac{\alpha}{2}}\, s\, \sqrt{p_{ii}}$$

and t is calculated for $n-(k+1)$ degrees of freedom.

$$s^2 = \frac{\sum\,(y-\hat{y})^2}{n-(k+1)}$$

We can construct a test of a hypothesis concerning a general linear model parameter a_i. The null hypothesis is $a_i = 0$. There are three research hypotheses ($a_i > 0$, $a_i < 0$, $a_i \neq 0$). We choose one depending on a particular experimental situation.

Test of an Hypothesis Concerning a_i

Research Hypothesis:
  1. $a_i > 0$
  2. $a_i < 0$
  3. $a_i \neq 0$

Null Hypothesis h_0: $a_i = 0$

Test Statistic:

$$t = \frac{\hat{a}_i}{s\,\sqrt{p_{ii}}}$$

Rejection Region: For a given value of α and the number of degrees of freedom $= n - k - 1$
  1. if $t > t_\alpha$ reject h_0

2.   if $t < -t_\alpha$ reject h_0

3.   if $t > \left| t_{\frac{\infty}{2}} \right|$ reject h_0

Estimates of E(y)

For a specific setting of

$$x_1, x_2, \ldots, x_k$$

we shall find the estimate of $E(y)$. Consider the prediction equation of that setting

$$\hat{y} = \hat{a}_0 + \hat{a}_1 x_1 + \ldots + \hat{a}_k x_k$$

Taking the repeated sampling at this specific setting we obtain the sampling distribution of $\hat{y}$ which has a mean

$$E(y) = a_0 + a_1 x_1 + \ldots + a_n x_n = P^T A$$

where $\quad P = \begin{bmatrix} 1 \\ x_1 \\ x_2 \\ \vdots \\ x_k \end{bmatrix} \quad$ and $\quad A = \begin{bmatrix} a_0 \\ a_1 \\ \vdots \\ a_k \end{bmatrix}$

Variance of the sampling distribution of $\hat{y}$ is

$$V(\hat{y}) = P^T (X^T X)^{-1} P \sigma^2$$

Assuming that the distribution of the random errors ε_i is normal we find

$$100 (1 - \alpha)\%$$

confidence interval for $E(y)$.

The value of t is calculated for $n - (k + 1)$ degrees of freedom.

$$\begin{array}{l} 100(1 - \alpha)\% \\ \text{Confidence Interval} \\ \text{for } E(y) \end{array} = \hat{y} \pm t_{\frac{\alpha}{2}} s \sqrt{P^T (X^T X)^{-1} P}$$

For a given setting of the independent variables we can design a statistical test of $E(y)$.

Research Hypothesis: 1. $E(y) > \mu_0$

    2. $E(y) < \mu_0$

    3. $E(y) \neq \mu_0$

Null Hypothesis h_0: $E(y) = \mu_0$

Test Statistic: $t = \dfrac{\hat{y} - \mu_o}{s \sqrt{P^T (X^T X)^{-1} P}}$

Rejection Region: 1.   if $t > t_\alpha$ reject h_0

    2.   if $t < t_\alpha$ reject h_0

    3.   if $t > \left| t_{\frac{\alpha}{2}} \right|$ reject h_0

12.3 CORRELATION

Regression or estimation enables us to estimate one variable (the dependent variable) from one or more independent variables.

Correlation establishes the degree of relationship between variables. It answers the question: how well a given equation describes or explains the relationship between independent and dependent variables.

Perfect Correlation

If all values of the variables fit the equation without errors we say that the variables are perfectly correlated.

The area of a square S is in perfect correlation to its side d

$$S = d^2$$

Tossing two coins we record the result for each coin. There is no relationship between the results for each coin, assuming the coins are fair, that is they are uncorrelated.

Between perfectly correlated and uncorrelated situations there are situations with some degree of correlation. The height and weight of people shows some correlation.

Simple correlation and simple regression occurs when only two variables are involved. When more than two variables are involved, we speak of multiple correlation.

Correlation Coefficient

The degree of relationship between two variables x and y is described by the correlation coefficient. If n observations are given

$$(x_i, y_i)\ i = 1, 2, ..., n$$

We can compute the sample correlation coefficient r

$$r = \frac{\sum (x - \overline{x})(y - \overline{y})}{\sqrt{\sum (x - \overline{x})^2 \sum (y - \overline{y})^2}}$$

We shall list some properties of r

1. $-1 < r < 1$

2. $r > 0$ indicates a positive linear relationship and $r < 0$ indicates a negative linear relationship.

3. $r = 0$ indicates no linear relationship.

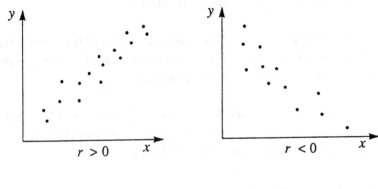

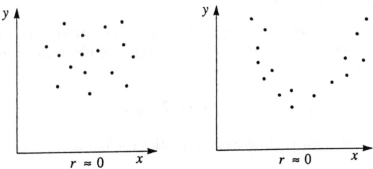

If y tends to increase as x increases the correlation is called positive.

Consider the linear model

$$y = a_0 + a_1 x + \varepsilon$$

The total variability of the y's about their mean $\bar{y}$ can be expressed as

$$\Sigma(y - \bar{y})^2 = \Sigma(y - \hat{y})^2 + \Sigma(\hat{y} - y)^2$$

$\Sigma(\hat{y} - \bar{y})^2$ is that portion of the total variability that can be accounted for by the independent variable x. We have

$$\Sigma(y - \bar{y})^2 = \Sigma y^2 - \frac{(\Sigma y)^2}{n}$$

$$\Sigma(y - \hat{y})^2 = \Sigma(y - \bar{y})^2 - \frac{(\Sigma(x - \bar{x})(y - \bar{y}))^2}{\Sigma(x - \bar{x})^2}$$

Hence

$$\Sigma(\hat{y} - \bar{y})^2 = \frac{(\Sigma(x - \bar{x})(y - \bar{y}))^2}{\Sigma(x - \bar{x})^2}$$

and

$$\frac{\Sigma(y - \hat{y})^2}{\Sigma(y - \bar{y})^2} = 1 - r^2$$

$$\frac{\Sigma(\hat{y} - \bar{y})^2}{\Sigma(y - \bar{y})^2} = r^2$$

The total variation of y defined as

$$\Sigma \ (y - \bar{y})^2$$

is the sum of

$$\Sigma(y - \overline{y})^2 = \Sigma(y^2 - \hat{y})^2 + \Sigma(\hat{y} - \overline{y})^2$$

where $\Sigma(y - \hat{y})^2$ is called the unexplained variation and $\Sigma(\hat{y} - \overline{y})^2$ is called the explained variation. The deviations $y - \hat{y}$ behave in a random manner, while $\hat{y} - \overline{y}$ have a definite pattern.

We define

$$\text{Coefficient of determination} = \frac{\text{explained variation}}{\text{total variation}}$$

The value of coefficient of determination lies between zero and one. Note that

$$r^2 = \text{coefficient of determination}$$

and

$$r = \pm \sqrt{\frac{\text{explained variation}}{\text{total variation}}}$$

Standard Error of Estimate

A set of measurements is given

$$(x_1, y_1)\,(x_2, y_2), \ldots, (x_n, y_n)$$

Suppose x is an independent variable. Then the least square regression line of y on x is

$$y = a_0 + a_1 x \tag{1}$$

Coefficients a_0 and a_1 can be computed from the normal equations

$$\left.\begin{array}{l} \Sigma y = a_0 n + a_1 \Sigma x \\ \Sigma xy = a_0 \Sigma x + a_1 \Sigma x^2 \end{array}\right\} \qquad (2)$$

$$a_0 = \frac{(\Sigma y)(\Sigma x^2) - (\Sigma x)(\Sigma xy)}{n\Sigma x^2 - (\Sigma x)^2}$$

$$a_1 = \frac{n\Sigma xy - (\Sigma x)(\Sigma y)}{n\Sigma x^2 - (\Sigma x)^2} \qquad (3)$$

If y is an independent variable then

$$x = b_0 + b_1 y \qquad (4)$$

and the system of normal equations is

$$\left.\begin{array}{l} \Sigma x = b_0 n + b_1 \Sigma y \\ \Sigma xy = b_0 \Sigma x + b_1 \Sigma y^2 \end{array}\right\} \qquad (5)$$

From (5) we find the coefficients

$$b_0 = \frac{(\Sigma x)(\Sigma y^2) - (\Sigma y)(\Sigma xy)}{n\Sigma y^2 - (\Sigma y)^2}$$

$$b_1 = \frac{n\Sigma xy - (\Sigma x)(\Sigma y)}{n\Sigma y^2 - (\Sigma y)^2} \qquad (6)$$

By $\hat{y}$ we denote the value of y for a given value of x computed from eq. (1).

We define the standard error of estimate of y on x by

$$s_y = \sqrt{\frac{\sum(y - \hat{y})^2}{n}} \tag{7}$$

On the other hand if eq. (4) is used we have

$$s_x = \sqrt{\frac{\sum(x - \hat{x})^2}{n}} \tag{8}$$

The values of $\hat{y}$ estimated from the regression line are given by

$$\hat{y} = a_0 - a_1 x \tag{9}$$

Hence (7) can be written as

$$S_y^2 = \frac{\sum(y - \hat{y})^2}{n} = \frac{\sum(y - a_0 - a_1 x)^2}{n} \tag{10}$$

$$= \frac{\sum y(y - a_0 - a_1 x) - a_0 \sum(y - a_0 - a_1 x) - a_0 \sum x(y - a_0 - a_1 x)}{n}$$

We can reduce (10) using

$$\sum(y - a_0 - a_1 x) = \sum y - a_0 n - a_1 \sum x = 0$$

$$\sum x(y - a_0 - a_1 x) = \sum xy - a_0 \sum x - a_1 \sum x^2 = 0$$

since

$$\sum xy = a_0 \sum x = a_1 \sum x^2$$

Eq. (10) becomes

$$S_y^2 = \frac{\sum y(y - a_0 - a_1 x)}{n} = \frac{\sum y^2 - a_0 \sum y - a_1 \sum xy}{n} \quad (11)$$

We have defined the total variation as

$$\text{total variation} = \sum(y - \bar{y})^2$$

and have shown that

Total Variation = Unexplained Variation + Explained Variation

$$\sum(y - \bar{y})^2 = \sum(y - \hat{y})^2 + \sum(\hat{y} - \bar{y})^2$$

The ratio

$$0 \leq \frac{\text{explained variation}}{\text{total variation}} \leq 1$$

is always non-negative. Thus, we have denoted it by r^2.

Coefficient of determination $= r^2 = \dfrac{\text{explained variation}}{\text{total variation}}$

$$= \frac{\sum(\hat{y} - \bar{y})^2}{\sum(y - \bar{y})^2}$$

Also we have defined

Coefficient of correlation $= r = \pm \sqrt{\dfrac{\text{explained variation}}{\text{total variation}}}$

$$= \pm \sqrt{\frac{\sum(\hat{y} - \bar{y})^2}{\sum(y - \bar{y})^2}}$$

Note that r is dimensionless and

$$-1 \leq r \leq 1$$

The sign $+$ is used for positive linear correlation and $-$ is used for negative linear correlation.

The standard deviation of y is

$$s = \sqrt{\frac{\sum(y - \bar{y})^2}{n}} \qquad (12)$$

Hence, using (10) we obtain

$$r = \pm \sqrt{\frac{\sum(\hat{y} - \bar{y})^2}{\sum(y - \bar{y})^2}} = \pm \sqrt{\frac{\sum(y - \bar{y})^2 - \sum(y - \hat{y})^2}{\sum(y - \bar{y})^2}} \qquad (13)$$

$$= \pm \sqrt{1 - \frac{s_y^2}{s^2}}$$

or

$$s_y = s\sqrt{1 - r^2} \qquad (14)$$

Definition

$$r = \pm \sqrt{\frac{\text{explained variation}}{\text{total variation}}}$$

is general and applies to linear as well as nonlinear relationships. For the non-linear case the values of $\hat{y}$ are computed from an applicable non-linear equation.

Suppose the situation is described by

$$y = a_0 + a_1 x + \ldots + a_k x^k \qquad (15)$$

Then S_y of eq. (11) becomes

$$s_y = \sqrt{\frac{\sum(y - \hat{y})^2}{n}}$$

$$= \sqrt{\frac{\sum y^2 - a_0 \sum y - a_1 \sum xy - \ldots - a_k \sum x^k y}{n}} \qquad (16)$$

The modified standard error of estimate is given by

$$\hat{s}_y = \sqrt{\frac{n}{n - k - 1}}\, S_y \qquad (17)$$

Here $n - k - 1$ is the number of degrees of freedom.

We shall briefly discuss how the coefficient of correlation works. Suppose experimental data are gathered, (x_1, y_1), ..., (x_n, y_n). Then we assume some kind of relationship between x and y. First choice is between

1. Linear and
2. Non-linear

If the relationship is non-linear we have to assume some non-linear equation relating x to y. Then, depending on what type of relationship we assume, the value of r is calculated.

One should keep in mind that r depends on what kind of equation is assumed.

Suppose we assume the linear relationship between x and y

and the computed value of r is close to zero. All it means is that there is no linear relationship between x and y.

Then, it is possible that:

1. There is no relationship between x and y.

2. There is a non-linear relationship between x and y.

12.4 COMPUTATIONAL FORMULAS

The least square regression line is

$$y = a_0 + a_1 x$$

can be written as

$$Y = \left(\frac{\sum XY}{\sum X^2} \right) X \qquad (18)$$

where

$$X = x - \bar{x}$$
$$Y = y - \bar{y} \qquad (19)$$

Observe that a least square line always passes through the point $(\bar{x}, \bar{y})$. If x is an independent variable

$$y = a_0 + a_1 x$$

and

$$\sum y = a_0 n + a_1 \sum x$$

Dividing both sides by n we find

$$\overline{y} = a_0 + a_1\overline{x} \qquad (20)$$

Hence

$$y - \overline{y} = a_1(x - \overline{x}) \qquad (21)$$

Now we shall prove (18).

For $X = x - \overline{x}$ and $Y = y - \overline{y}$ eq. (21) becomes

$$Y = a_1 X \qquad (22)$$

The coefficients are

$$
\begin{aligned}
a_1 &= \frac{n\sum xy - (\sum x)(\sum y)}{n\sum x^2 - (\sum x)^2} \\[2mm]
&= \frac{n\sum(X + \overline{x})(Y + \overline{y}) - \sum(X + \overline{x})\sum(Y + \overline{y})}{n\sum(\overline{x} + X)^2 - (\sum(X + \overline{x}))^2} \qquad (23) \\[2mm]
&= \frac{n\sum XY + n\overline{y}\sum X + n\overline{x}\sum Y + n^2\overline{x}\,\overline{y} - (\sum X + n\overline{x})(\sum Y + n\overline{y})}{n\sum X^2 + 2n\overline{x}\sum X + n^2\overline{x}^2 - (\sum X + n\overline{x})^2}
\end{aligned}
$$

But

$$\sum X = \sum(x - \overline{x}) = 0$$
$$\sum Y = 0$$

Hence

$$a_0 = \frac{n\sum XY + n^2\overline{x}\,\overline{y} - n^2\overline{x}\,\overline{y}}{n\sum X^2 + n^2\overline{x}^2 - n^2\overline{x}^2} = \frac{\sum XY}{\sum X^2} \qquad (24)$$

The least square line is

$$Y = \left(\frac{\sum XY}{\sum X^2} \right) X \tag{25}$$

For the linear relationship between x and y the coefficient of correlation is

$$r = \pm \sqrt{\frac{\sum (\hat{y} - \bar{y})^2}{\sum (y - \bar{y})^2}} \tag{26}$$

The least square regression line of y on x can be written

$$\hat{y} = a_0 + a_1 x$$

or

$$\hat{Y} = a_1 X$$

where

$$a_1 = \frac{\sum XY}{\sum X^2} \qquad \text{and} \qquad \hat{Y} = \hat{y} - \bar{y}$$

Thus

$$r^2 = \frac{\sum (\hat{y} - \bar{y})^2}{\sum (y - \bar{y})^2} = \frac{\sum \hat{Y}^2}{\sum Y^2} = \frac{\sum a_1^2 X^2}{\sum Y^2}$$

$$= \frac{a_1^2 \sum X^2}{\sum Y^2} = \left(\frac{\sum XY}{\sum X^2} \right)^2 \cdot \frac{\sum X^2}{\sum Y^2} \tag{27}$$

$$= \frac{(\sum XY)^2}{(\sum X^2)(\sum Y^2)}$$

and

$$r = \pm \frac{\sum XY}{\sqrt{(\sum X^2)(\sum Y^2)}} \qquad (28)$$

Observe that for positive linear correlation $\hat{Y}$ increases as X increases and the value of

$$\frac{\sum XY}{\sqrt{(\sum X^2)(\sum Y^2)}} \qquad (29)$$

is positive. Similarly for negative linear correlation the value of (29) is negative. We can define the coefficient of linear correlation as

$$r = \frac{\sum XY}{\sqrt{(\sum X^2)(\sum Y^2)}} \qquad (30)$$

which automatically assumes the correct sign. Formula (30) is called the product-moment formula for the coefficient of linear correlation.

EXAMPLE:

The data collected are shown in the table below.

x	1	2	4	7	8	10	11
y	1	3	3	5	7	9	9

We want to determine if the relationship between x and y is of linear type. For that purpose the value of r has to be calculated.

x	y	$X = x - \bar{x}$	$Y = y - \bar{y}$	X^2	XY	Y^2
1	1	−5.143	−4.286	26.450	22.043	18.360
2	3	−4.143	−2.286	17.164	9.471	5.226
4	3	−2.143	−2.286	4.592	4.899	5.226
7	5	0.857	−0.286	0.734	−0.245	0.082
8	7	1.857	1.714	3.448	3.183	2.938
10	9	3.857	3.714	14.876	14.325	13.794
11	9	4.857	3.714	23.59	18.039	13.794
$\sum x$ =43	$\sum y$ =37			$\sum X^2 =$ 90.86	$\sum XY =$ 71.71	$\sum Y^2 =$ 59.43

$$\bar{x} = 6.143$$

$$\bar{y} = 5.286$$

$$r = \frac{\sum XY}{\sqrt{(\sum X^2)(\sum Y^2)}} = \frac{71.71}{\sqrt{90.86 \cdot 59.43}} = 0.976$$

There is strong positive linear correlation between x and y.

Covariance

The product-moment formula states that

$$r = \frac{\sum XY}{\sqrt{(\sum X^2)(\sum Y^2)}} \qquad (31)$$

The standard deviations of the variables x and y are

$$s_x = \sqrt{\frac{\sum X^2}{n}}$$

$$s_y = \sqrt{\frac{\sum Y^2}{n}} \tag{32}$$

We define the covariance of x and y as

$$s_{xy} = \frac{\sum XY}{n} \tag{33}$$

The coefficient of correlation can be written as

$$r = \frac{s_{xy}}{s_x s_y} \tag{34}$$

We shall derive another formula for the linear correlation coefficient.

$$X = x - \bar{x}$$

$$Y = y - \bar{y}$$

$$r = \frac{\sum XY}{\sqrt{(\sum X^2)(\sum Y^2)}} = \frac{\sum(x - \bar{x})(y - \bar{y})}{\sqrt{(\sum(x - \bar{x})^2)(\sum(y - \bar{y})^2)}} \tag{35}$$

But

$$\sum(x - \bar{x})(y - \bar{y}) = \sum xy - \frac{(\sum x)(\sum y)}{n} \tag{36}$$

and

$$\sum(x - \bar{x})^2 = \sum x^2 - \frac{(\sum x)^2}{n} \tag{37}$$

$$\sum(y - \bar{y})^2 = \sum y^2 - \frac{(\sum y)^2}{n} \tag{38}$$

Substituting (36), (37) and (38) into (35) we obtain

$$r = \frac{n\sum xy - (\sum x)(\sum y)}{\sqrt{(n\sum x^2 - (\sum x)^2)(n\sum y^2 - (\sum y)^2)}} \qquad (39)$$

Correlation Coefficient for Grouped Data

For grouped data the values of variables x and y coincide with the corresponding class marks and the frequencies f_x and f_y are the corresponding class frequencies. By f we denote the cell frequencies corresponding to the pairs of class marks (x, y).

EXAMPLE:

The weights and ages of a sample of students were measured. The results are shown in the table

Age	Weight 120-129	130-139	140-149
16-18	2	3	1
19-21	1	5	0
22-24	3	6	2
25-28	2	5	4

cell frequency

Consider the shaded cell. The class marks of this cell are

$$\text{weight} = 134.5$$

$$\text{age} = 20$$

The cell frequency of the cell (134.5, 20) is 5.

Formula (39) becomes

$$r = \frac{n\sum fxy - (\sum f_x x)(\sum fy)}{\sqrt{[n\sum f_x x^2 - (\sum f_x x)^2][n\sum f_y y^2 - (\sum f_y y)^2]}} \quad (40)$$

If the class interval widths are constant, say d_x and d_y then

$$x = A + d_x u_x$$
$$y = B + d_y u_y \quad (41)$$

and (40) can be written in the form

$$r = \frac{n\sum f u_x u_y - (\sum f_x u_x)(\sum f_y u_y)}{\sqrt{[n\sum f_x u_x^2 - (\sum f_x u_x)^2][n\sum f_y u_y^2 - (\sum f_y u_y)^2]}} \quad (42)$$

This method is called the cooling method.

Linear Correlation Coefficient

Regression line of y on x is

$$y = a_0 + a_1 x$$

or

$$y - \overline{y} = \frac{r s_y}{s_x}(x - \overline{x}) \quad (43)$$

or

$$Y = \frac{r s_y}{s_x} X$$

If we consider y to be an independent variable then

$$x = b_0 + b_1 y$$

or

$$x - \overline{x} = \frac{r s_x}{s_y}(y - \overline{y}) \quad (44)$$

or

$$X = \frac{rs_x}{s_y} Y$$

If $r = \pm 1$ two lines (43) and (44) are identical and there is perfect linear correlation between x and y. The slopes of lines (43) and (44) are equal if and only if

$$r = \pm 1.$$

If $r = 0$, then there is no linear correlation and the lines are at right angles.

Spearman's Formula for Rank Correlation

Sometimes the exact values of the variables are impossible or impractical to calculate. In such cases we use the method of ranking the data according to their importance, size, quality, etc. using the numbers $1, 2, \ldots, n$. For two variables x and y ranked that way we define the coefficient of rank correlation

$$r = 1 - \frac{6\sum d^2}{n(n^2 - 1)}$$

where n is the number of all pairs (x, y) of ranked data and d is the difference between ranks of corresponding values of x and y.

Sampling

Suppose the experimental situation requires the measurement of two variables x and y. The results are assembled in pairs

$$(x_1, y_1), \ldots, (x_n, y_n).$$

These pairs constitute a sample from a population of all possible pairs. Populations which involve two variables are called the bivariate populations. We can compute the coefficient of correlation r for the sample $(x_1, y_1), \ldots, (x_n, y_n)$. Based on it we can make estimates concerning the coefficient of correlation R for the whole population.

To test the significance or hypotheses of various values of R, we have to know the sampling distribution of r.

Test of Hypothesis R = 0

For $R = 0$ this distribution is symmetric and we can use a statistic involving Student's distribution.

The statistic

$$t = \frac{r\sqrt{n-2}}{\sqrt{1-r^2}}$$

has Student's distribution for $n - 2$ degrees of freedom.

Test of Hypothesis R = R₀ ≠ 0

For $R \neq 0$ the sampling distribution is skewed. The Fisher's Z transformation yields a statistic which is very close to normal distribution:

$$Z = \frac{1}{2} \ln\left(\frac{1+r}{1-r}\right) = 1.1513 \log_{10}\left(\frac{1+r}{1-r}\right)$$

The statistic Z has mean

$$u_z = \frac{1}{2} \ln\left(\frac{1+R_0}{1-R_0}\right) = 1.1513 \log_{10}\left(\frac{1+R_0}{1-R_0}\right)$$

and standard deviation given by

$$\sigma_2 = \frac{1}{\sqrt{n-3}}$$

Suppose two samples of sizes n_1 and n_2 are drawn from the population. The correlation coefficient for samples n_1 and n_2 are r_1 and r_2 respectively.

Using r_1 and r_2 we compute Z_1 and Z_2

$$Z_1 = \frac{1}{2} \ln \left(\frac{1 + r_1}{1 - r_1} \right)$$

$$Z_2 = \frac{1}{2} \ln \left(\frac{1 + r_2}{1 - r_2} \right)$$

The test statistic

$$Z = \frac{Z_1 - Z_2 - \mu_{Z_1 - Z_2}}{\sigma_{Z_1 - Z_2}}$$

where

$$\mu_{Z_1 - Z_2} = \mu_{Z_1} - \mu_{Z_2}$$

and

$$\sigma_{Z_1 - Z_2} = \sqrt{\sigma^2_{Z_1} + \sigma^2_{Z_2}} = \sqrt{\frac{1}{n_1 - 3} + \frac{1}{n_2 - 3}}$$

has normal distribution. Thus, we can decide if the two correlation coefficients r_1 and r_2 differ significantly from each other.

CHAPTER 13

EXPERIMENTAL DESIGN

13.1 DESIGN OF AN EXPERIMENT

A statistician starts his job with the design of an experiment. Depending on the experiment, he needs a certain amount of information. The cost of gathering information should also be considered.

The design of an experiment should be cost effective and should guarantee that the necessary information will be supplied.

Each design of an experiment consists of the following steps:

1. Statement of an objective. It usually consists of a description of the population and a description of the parameters of the population we have to estimate.

2. Statement of the amount of information required about the parameters.

3. Experimental design which consists of selection of the appropriate experimental plan.

4. Estimation of test procedure.

Suppose the objective is to find the parameter μ of the population. We have to decide how accurate μ should be. For that purpose we can apply a bound on the error of estimate of μ. That bound can be $\pm A$ units of μ from μ.

Note that A can assume any value we choose.

Suppose μ is a mean yearly income of a construction worker. Then we can choose $A = \$1.00$ and estimate μ to be within $\$1.00$.

Often the sample mean is used as a point estimate of μ; then, the bound on the error of estimate is

$$\frac{2\sigma}{\sqrt{n}}$$

Solving the equation

$$A = \frac{2\sigma}{\sqrt{n}}$$

for n

$$n = \frac{u\sigma^2}{A^2}$$

we find the sample size required to estimate μ to within A units.

Two terms are used frequently in descriptions of experimental design.

Experimental Unit

Any object or person on which a measurement is made is called an experimental unit

Treatment

A treatment is a factor level, or combination of factor levels, applied to an experimental unit.

Suppose our experimental objective is to estimate the parameters μ and $\mu_1 - \mu_2$ or to test a hypothesis about them.

It is important to determine the quantity of information in an experiment relative to the parameter.

Often we estimate a parameter λ, using a point estimate $\hat{\lambda}$ For the sampling distribution of the point estimates approximately normal with mean $\hat{\lambda}$ and standard deviation $\sigma_{\hat{\lambda}}$

$$\text{the bound on the error of estimate} = 2\sigma_{\hat{\lambda}} \ .$$

If we are looking for a point estimate of λ then we can set the bound on the error of estimate, say A.

In general

$$\sigma_{\hat{\lambda}} = \sigma_{\hat{\lambda}}(n)$$

that is $\sigma_{\hat{\lambda}}$ depends on n. Hence solving the equation

$$A = 2\sigma_{\hat{\lambda}}$$

for n we find the sample size n necessary to achieve the bound on the error equal A.

We summarize the results concerning sample sizes in the table

λ	$\hat{\lambda}$	$\sigma_{\hat{\lambda}}$	Sample Size
μ	$\overline{y}_1$	$\dfrac{\sigma}{\sqrt{n}}$	$n = \dfrac{\mu\sigma^2}{A^2}$
$\mu_1 - \mu_2$	$\overline{y}_1 - \overline{y}_2$	$\sqrt{\dfrac{\sigma_1^2}{n} + \dfrac{\sigma_2^2}{n}}$	$n = \dfrac{\mu(\sigma_1^2 + \sigma_2^2)}{A^2}$

Note that in estimating

$$\mu_1 - \mu_2$$

we use

$$\bar{y}_1 - \bar{y}_2$$

and determine the sample size by solving the equation

$$A = 2\sqrt{\frac{\sigma_1^2}{n} + \frac{\sigma_2^2}{n}}$$

We want to estimate $\mu_1 - \mu_2$ to within A units.

EXAMPLE:

A chemist wants to estimate the difference in mean melting temperatures for two different alloys. From previous experiments he knows that the range in melting temperatures for each alloy is approximately 400°. How many independent random samples of each kind of alloy must be checked to estimate

$$T_1 - T_2$$

within 50°?

The range of melting temperatures for both alloys is the same, thus we can assume that the population variances σ_1^2 and σ_2^2 are approximately the same.

$$\sigma_1^2 = \sigma_2^2 = \sigma^2$$

The range estimate of σ is

$$\hat{\sigma} = \frac{\text{range}}{4} = \frac{400}{4} = 100$$

The sample size formula

$$n = \frac{4(\sigma_1^2 + \sigma_2^2)}{A^2}$$

$$= \frac{4(100^2 + 100^2)}{50^2} = 32$$

Thus we should examine 32 samples of alloy of each type to estimate $T_1 - T_2$ with a bound on the error of estimate

$$A = 50°$$

Observe that increasing the desired bound on the error of estimate we decrease the number of required samples, i.e., decrease the cost of conducting the experiment.

For example for

$$A = 75°$$

$$n = \frac{4(100^2 + 100^2)}{75^2} = 14$$

Now, suppose the objective is to find an interval estimate of the parameter λ. Let $\hat{\lambda}$ be a point estimate of the parameter λ. We assume that the sampling distribution of the point estimates is approximately normal, with mean λ and standard deviation $\sigma_{\hat{\lambda}}$.

For confidence coefficient

$$1 - \alpha$$

the confidence interval for λ is

$$\hat{\lambda} \pm Z_{\frac{\alpha}{2}} \sigma_{\hat{\lambda}}$$

One of the accepted measures of the amount of information important to the parameter λ is the half width $Z_{\frac{\alpha}{2}} \sigma_{\hat{\lambda}}$ of the confidence interval.

$$A = Z_{\frac{\alpha}{2}} \sigma_{\hat{\lambda}}$$

Solving this equation for n we find the sample size required to estimate a parameter λ by using confidence coefficient

$$1 - \alpha$$

and a confidence interval $\pm Z_{\frac{\alpha}{2}} \sigma_{\hat{\lambda}}$

The table below shows the results of interval estimates of the parameters μ and $\mu_1 - \mu_2$.

λ	$\hat{\lambda}$	Confidence Interval	Sample Size
μ	$\overline{y}$	$\overline{y} \pm Z_{\frac{\alpha}{2}} \dfrac{\sigma}{\sqrt{n}}$	$n = \dfrac{z_{\frac{\alpha}{2}}^2 \sigma^2}{A^2}$
$\mu_1 - \mu_2$	$\overline{y}_1 - \overline{y}_2$	$(\overline{y}_1 - \overline{y}_2) \pm Z_{\frac{\alpha}{2}} \sqrt{\dfrac{\sigma_1^2}{n} + \dfrac{\sigma_2^2}{n}}$	$n = \dfrac{z_{\frac{\alpha}{2}}^2 (\sigma_1^2 + \sigma_2^2)}{A^2}$

Suppose we want to test the research hypothesis

$$h_a : \lambda > \lambda_0$$

Then the null hypothesis is

$$h_0 : \lambda = \lambda_0$$

Assume that the distribution of $\hat{\lambda}$ is approximately normal with mean λ_0 and standard deviation $\sigma_{\hat{\lambda}}$ under the null hypothesis. We want the probability of a type 1 error to be α and the probability of a type II error to be β or less for the actual value of λ such that

$$\lambda - \lambda_0 \geq \Delta$$

i.e., λ lies a distance of Δ or more above λ_0.

λ	h_0	Δ	Test Statistic	Sample Size
μ	$\mu = \mu_0$	$\lvert \mu - \mu_0 \rvert$	$Z = \dfrac{\overline{y} - \mu_0}{\sigma}\sqrt{n}$	$n = \dfrac{\sigma^2 (Z_\alpha + Z_\beta)^2}{\Delta^2}$
$\mu_1 - \mu_2$	$\mu_1 - \mu_2 = \delta$	$\lvert \mu_1 - \mu_2 - \delta \rvert$	$Z = \dfrac{(\overline{y}_1 - \overline{y}_2) - \delta}{\sqrt{\dfrac{\sigma_1^2}{n} + \dfrac{\sigma_2^2}{n}}}$	$n = \dfrac{(\sigma_1^2 + \sigma_2^2)(Z_\alpha + Z_\beta)^2}{\Delta^2}$

13.2 COMPLETELY RANDOMIZED DESIGN

Sometimes we deal with more than one population. The methods of estimating a population mean μ or the difference between two population means $\mu_1 - \mu_2$ described in 13.1 have to be extended. The completely randomized design enables us to compare u ($u \geq 2$) population means,

$$\mu_1, \ \mu_2, \ \ldots, \mu_u.$$

We assume here that there are u different populations. From these populations we draw independent random samples of size

$$n_1, n_2, \ldots, n_u$$

respectively.

Thus, we assume that there are

$$n_1 + n_2 + \ldots + n_u$$

experimental units. Each unit receives a treatment. Treatments are randomly assigned to the experimental units in such a way that

n_1 units receive treatment 1

n_2 units receive treatment 2

.

.

.

n_u units receive treatment u

After the corresponding treatment means are calculated we will make inferences concerning them.

EXAMPLE:

A factory has four different methods at its disposal to test the quality and durability of the bearings it makes. It is important to determine if there is a difference in mean test readings for bearings using four different methods.

Here the experimental units are bearings and the treatments are four methods of testing A, B, C, D. A random sample of 16 bearings is chosen and 4 bearings are randomly assigned to each testing method. We have a completely randomized design with four observations for each treatment.

	Bearing			
Testing Method	A	B	C	D
	A	B	C	D
	A	B	C	D
	A	B	C	D

Suppose four technicians are assigned to perform tests. We denote them 1, 2, 3, 4. The bearings are randomly assigned, four to each technician.

One of the possible random assignments is shown in the table.

Technician

1	2	3	4
A	B	C	D
A	B	C	D
A	B	C	D
A	B	C	D

For each treatment we have four observations. Now any detected difference may be due to

1. differences among methods.

2. differences among technicians.

Suppose that the hypothesis

$$h_0 : \mu_A - \mu_B = 0$$

was tested against

$$h_a : \mu_A - \mu_B \neq 0$$

The rejection of h_0 could be due to the differences in the methods of testing or due to the fact that technicians represent various skill levels.

The completely randomized design presented above can be

modified to gain more reliable information about the means μ_A, μ_B, μ_C, μ_D. We can impose one restriction upon the random choice of methods of testing.

Each technician can be asked to use each of the methods exactly once. The order of tests for each technician is randomized.

One of the possible designs is shown in the table.

Technician

1	2	3	4
B	C	A	A
D	A	D	D
A	D	B	C
C	B	C	B

This design which is an improved completely randomized design, is called a randomized block design. Here the blocks are technicians. The influence of technicians upon the results μ_A, μ_B, μ_C, μ_D in this design has been eliminated.

Now, under the hypothesis

$$h_0 : \mu_A - \mu_B = 0$$

is rejected we know that the difference between μ_A and μ_B is due to the differences detected by the methods A and B.

EXAMPLE:

The situation is as described in the previous example. The instruments used to test the bearings require careful calibration and adjustments. Each of the technicians is able to perform only one test a day.

We use the randomized block design table and denote the first row as a workload of day 1, the second row as a workload of day 2, etc.

Technician

		1	2	3	4
Day	1	B	C	A	A
	2	D	A	D	D
	3	A	D	B	C
	4	C	B	C	B

Now, the possible differences in μ_A, μ_B, μ_C, μ_D can be due to the fact that no D method was used on day 1 while 3 D methods were used on day 2. Hence, if we reject

$$h_0: \mu_B - \mu_D = 0$$

We would not be certain if the difference between μ_B and μ_D is due to the difference in testing methods B and D or due to the difference among days.

In this situation the experimental units are affected by two foreign sources of variability: technicians and days.

We can take care of this by using the Latin square design. The completely randomized design was modified to eliminate the technician as an extraneous source of variability. The obtained design was called the randomized block design.

We can further modify this design to eliminate the influence of days, that is the variability among days. The randomization is restricted to guarantee that each method appears in each row and in each column. Such experimental design is called a Latin Square design.

One of possible Latin Square designs is shown in the table.

Technician

		1	2	3	4
Day	1	B	C	A	D
	2	D	A	B	C
	3	C	B	D	A
	4	A	D	C	B

Here each method of testing is used once a day and once by each of the technicians. This design enables us to use pairwise comparisons of the testing procedures.

A Latin Square design has its limitations. It is used to compare u treatment means when there are at most two extraneous sources of variability present. The influence of these sources is eliminated by special arrangement into u rows and u columns. The u treatments are randomly assigned to the rows and columns in such a way that each treatment appears in every row and every column of the Latin Square design.

CHAPTER 14

COUNT DATA.
THE CHI-SQUARE TEST

14.1 COUNT DATA

In many situations we gather the sample data measured on a quantitative scale. For example to find the average yearly income of an American family we choose a sample of families and measure their incomes. Number of dollars is the result of each such measurement.

On the other hand, in some situations, the measurements are not quantitative. For example, if we want to classify students into three categories: good, average, bad. Data obtained from such experiments are called count data. We will be using the concept of a multinomial experiment.

Definition of Multinomial Experiment

1. The experiment consists of n identical trials.

2. There are k possible outcomes. Each trial results is one of k outcomes.

3. By p_i, $i = 1, ..., k$ we denote the probability that the outcome of a single trial will be i. This probability remains constant

from trial to trial.

4. The outcome of one trial does not have any influence upon another trial.

5. The total number of trials resulting in outcome i is n_i. We have

$$\sum_{i=1}^{k} n_i = n$$

Observe that since there are k outcomes

$$\sum_{i=1}^{k} p_i = 1$$

The multinomial distribution is the probability distribution for the number of observations resulting in each of the k outcomes. It is given by the formula

$$P(n_1, n_2, \ldots, n_k) = \frac{n!}{n_1!\, n_2! \ldots n_k!}\, p_1^{n_1} p_2^{n_2} \ldots p_k^{n_k}$$

14.2 DEFINITION OF χ^2

Let

$$E_1, \ldots, E_k$$

denote the set of possible events. The observed frequencies of events are

$$f_1, f_2, \ldots, f_k$$

respectively. Having the theoretical probabilities of events we can compute the theoretical or expected frequencies of events

$$e_1, \ldots, e_k$$

respectively.

It is sometimes crucial to find out if the theoretical frequencies differ significantly from the observed frequencies.

For that purpose we define the statistic χ^2

$$\chi^2 = \frac{(f_1 - e_1)^2}{e_1} + \frac{(f_2 - e_2)^2}{e_2} + \ldots + \frac{(f_k - e_k)^2}{e_k} \qquad (1)$$

χ^2 measures the discrepancy between observed and expected frequencies.

If the total frequency is n

$$\sum_{i=1}^{k} f_i = \sum_{i=1}^{k} e_i = n$$

Formula (1) can be written as

$$\chi^2 = \sum_{i=1}^{k} \frac{f_i^2}{e_i} - n \qquad (2)$$

Indeed

$$\chi^2 = \sum \frac{(f_i - e_i)^2}{e_i} = \sum \left(\frac{f_i^2 - 2f_i e_i + e_i^2}{e_i} \right)$$

$$= \sum \frac{f_i^2}{e_i} - 2\sum f_i + \sum e_i = \sum \frac{f_i^2}{e_i} - n$$

When $\chi^2 = 0$, then observed and expected frequencies agree exactly.

The sampling distribution of χ^2 is approximated closely by the χ^2 – distribution

$$y = y_0 \, x^{v-2} e^{-\frac{1}{2}\chi^2} \tag{3}$$

where v is the number of degrees of freedom.

Two situations occur:

1. The theoretical frequencies can be computed without the use of the population parameters estimated from sample statistics. Since

$$\sum_{i=1}^{k} e_i = n$$

we have to know only $k - 1$ of the expected frequencies.

$$v = k - 1$$

2. The theoretical frequencies can be computed only by m population parameters estimated from sample statistics

$$v = k - 1 - m$$

Using the null hypothesis h_0 we compute the expected frequencies and then the value of χ^2. If the value of χ^2 is greater than some critical value, we conclude that observed frequencies differ significantly from expected frequencies and reject h_0.

Otherwise, we accept h_0.

This procedure is called the chi-square test of hypothesis.

EXAMPLE:

A coin was tossed 500 times. The result was 283 heads and 217 tails. Using the level of significance of 0.01 check the hypothesis that the coin was fair.

Observed frequencies of heads and tails are respectively

$$f_1 = 283$$
$$f_2 = 217$$

and expected frequencies are

$$e_1 = 250$$
$$e_2 = 250$$

We have

$$\chi^2 = \frac{(f_1 - e_1)^2}{e_1} + \frac{(f_2 - e_2)^2}{e_2} = 4.356 + 4.356$$

$$= 8.712$$

There are two categories: heads and tails. Hence

$$k = 2$$

and the number of degrees of freedom is

$$\nu = k - 1 = 1$$

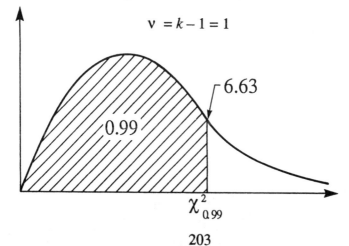

For $\nu = 1$ the value of $\chi^2_{0.99}$ is 6.63

$$\chi^2_{0.99} = 6.63$$

The value of χ^2 is 8.712. Since

$$8.712 > 6.63$$

we reject the hypothesis that the coin is fair at a 0.01 level of significance.

Observe that the chi-square distribution is a nonsymmetrical distribution.

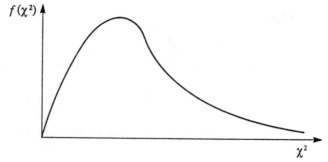

There are many chi-square distributions. We choose the appropriate one depending on the number of degrees of freedom.

The chi-square test is often used to determine the goodness of fit, that is to determine how well theoretical distributions (such as binomial, normal, etc.) fit empirical distributions.

14.3 CONTINGENCY TABLES

Let

$$E_1, E_2, \ldots E_k$$

be the set of possible events. The observed frequencies of events are

$$f_1, \ldots, f_k$$

correspondingly. The theoretical or expected frequencies are

$$e_1, \ldots, e_k$$

Event	E_1	E_2	. . .	E_k
Observed frequency	f_1	f_2	. . .	f_k
Expected frequency	e_1	e_2	. . .	e_k

In the table above observed frequencies occupy one row. Such a table is called a one-way classification table. The number of columns is k, so the table is

$$1 \times k \text{ table.}$$

There are tables where the observed frequencies occupy h rows. Such tables are called

$$h \times k \text{ tables.}$$

In general these tables are referred to as contingency tables. To each observed frequency in an $h \times k$ table there corresponds a theoretical frequency, usually computed according to the rules of probability. The contingency table consists of the cells.

Each cell contains a frequency called cell frequency. The total frequency in each row or each column is called the marginal frequency.

The statistic

$$\chi^2 = \frac{\Sigma(f_i - e_i)^2}{e_i} \qquad (4)$$

is applied to check agreement between observed and expected frequencies. The sum is taken over all cells in the contingency table. For h x k contingency tables this sum contains $h \cdot k$ terms. The sum of all observed frequencies is equal to the sum of all expected frequencies

$$\Sigma f_i = \Sigma e_i = n \qquad (5)$$

The statistic χ^2 defined by (4) has a sampling distribution approximated very closely by

$$y = y_0 \chi^{v-2} e^{-\frac{1}{2}\chi^2} \qquad (6)$$

Approximation gets better for large expected frequencies. The number of degrees of freedom v for

$$h > 1 , \, k > 1$$

should be calculated as follows:

1. When the expected frequencies are computed without using population parameters estimated from sample statistics

$$v = (h - 1)(k - 1) \qquad (7)$$

2. When m estimated parameters are necessary to compute the

expected frequencies

$$\nu = (h - 1)(k - 1) - m \qquad (8)$$

Contingency tables can be extended to higher dimensions. For example we can design a table

$$h \times k \times l$$

where three classifications are present.

EXAMPLE:

Consider the 2 x 2 contingency table

	I	II	Total
A	a_1	a_2	n_a
B	b_1	b_2	n_b
Total	n_1	n_2	n

Results Observed

Under a null hypothesis we find the expected frequencies

	I	II	Total
A	$\dfrac{n_1 n_a}{n}$	$\dfrac{n_2 n_a}{n}$	n_a
B	$\dfrac{n_1 n_b}{n}$	$\dfrac{n_2 n_b}{n}$	n_b
Total	n_1	n_2	n

Results Expected

Having observed and expected frequencies we can compute χ^2

$$\chi^2 = \frac{\left(a_1 - \frac{n_1 n_a}{n}\right)^2}{\frac{n_1 n_a}{n}} + \frac{\left(a_2 - \frac{n_2 n_a}{n}\right)^2}{\frac{n_2 n_a}{n}}$$

$$+ \frac{\left(b_1 - \frac{n_1 n_b}{n}\right)^2}{\frac{n_1 n_b}{n}} + \frac{\left(b_2 - \frac{n_2 n_b}{n}\right)^2}{\frac{n_2 n_b}{n}} \tag{9}$$

But

$$a_1 - \frac{n_1 n_a}{n} = a_1 - \frac{(a_1 + b_1)(a_1 + a_2)}{a_1 + a_2 + b_1 + b_2} = \frac{a_1 b_2 - a_2 b_1}{n}$$

Similarly we find that

$$a_2 - \frac{n_2 n_a}{n} = \frac{a_1 b_2 - a_2 b_1}{n}$$

etc.

$$\chi^2 = \frac{n}{n_1 n_a}\left(\frac{a_1 b_2 - a_2 b_1}{n}\right)^2 + \frac{n}{n_2 n_a}\left(\frac{a_1 b_2 - a_2 b_1}{n}\right)^2$$

$$+ \frac{n}{n_1 n_b}\left(\frac{a_1 b_2 - a_2 b_1}{n}\right)^2 + \frac{n}{n_2 n_b}\left(\frac{a_1 b_2 - a_2 b_1}{n}\right)^2 \tag{10}$$

$$= \frac{n(a_1 b_2 - a_2 b_1)^2}{n_a n_b n_1 n_2}$$

EXAMPLE:

New serum is tested for its effectiveness. Two samples A and B are tested each consisting of 100 sick patients. Group A used serum and Group B did not use serum.

	Recovered	Did Not Recover	Total
A (serum)	80	20	100
B (no serum)	60	40	100
Total	140	60	200

Frequencies Observed

The results are shown in the table. The null hypothesis h_0 states that serum has no effect.

We calculate the expected frequencies under h_0 null hypothesis.

	Recovered	Did Not Recover	Total
A (serum)	70	30	100
B (no serum)	70	30	100
Total	140	60	200

Frequencies Expected under h_0

We have

$$\chi^2 = \frac{(80-70)^2}{70} + \frac{(60-70)^2}{70} + \frac{(20-30)^2}{30} + \frac{(40-30)^2}{30}$$

$$= 9.524$$

Using the formula

$$v = (h - 1)(k - 1)$$

we find the number of degrees of freedom

$$v = 1$$

Our results are $\chi^2 = 9.524$ for $v = 1$. For one degree of freedom

$$\chi^2_{0.995} = 7.88$$

The results are significant even at 0.005 level

$$7.88 < 9.524$$

We reject h_0 and conclude that the serum is effective.

Coefficient of Contingency

The degree of dependence or relationship of the classifications in a contingency table is measured by the coefficient of contingency defined as

$$\lambda = \sqrt{\frac{\chi^2}{\chi^2 + n}} \qquad (11)$$

The degree of dependence increases with the value of λ. Note that

$$0 \leq \lambda < 1$$

If the number of rows and columns in the contingency table is equal to h the maximum value of λ is

$$\lambda_{max} = \sqrt{\frac{h - 1}{h}}$$

EXAMPLE:

For the last example the value of χ^2 was

$$\chi^2 = 9.524$$

and

$$n = 200$$

The value of the coefficient of contingency is

$$\lambda = \sqrt{\frac{9.524}{9.524 + 200}} = 0.213$$

Additive Property of χ^2

Repeated experiments are performed. For each one the value of χ^2 and ν is calculated.

$$\chi_1^2, \ \chi_2^2, \ \chi_3^2, \ \cdots$$

$$\nu_1, \ \nu_2, \ \nu_3, \ \cdots$$

The results of all these experiments are equivalent to one experiment with χ^2 given by

$$\chi^2 = \chi_1^2 + \chi_2^2 + \chi_3^2 + \ \cdots$$

and the number of degrees of freedom

$$\nu = \nu_1 + \ \nu_2 + \ \nu_3 + \ \cdots$$

EXAMPLE:

An experiment is performed four times in order to test hypothesis h_0. The values of χ^2 were

$$1.98, \ \ 3.14, \ \ 2.37, \ \ 1.90$$

In each case $v = 1$. At the level 0.05 for one degree of freedom, $v = 1$

$$\chi^2_{0.95} = 3.84$$

Thus we cannot reject h_0 on the basis of any one experiment. Combining the results of the four experiments we find

$$\chi^2 = 1.98 + 3.14 + 2.37 + 1.90 = 9.39$$

$$v = 1 + 1 + 1 + 1 = 4$$

for $v = 4$

$$\chi^2_{0.95} = 9.49$$

So we can reject h_0 at the 0.05 level of significance.

14.4 YATES' CORRECTION. SOME HELPFUL FORMULAS

Often we apply the equations for continuous distributions to the sets of discrete data.

For the chi-square distribution we use Yates' correction defined by

$$\chi^2_{corrected} = \frac{(|f_1 - e_1| - 0.5)^2}{e_1} + \dots + \frac{(|f_k - e_k| - 0.5)^2}{e_k}$$

The correction is made only when the number of degrees of freedom is one, $v = 1$.

Finally we show how to compute χ^2 which involve only the observed frequencies.

2 x 2 Tables

	I	II	Total
A	a_1	a_2	n_a
B	b_1	b_2	n_b
Total	n_1	n_2	n

$$\chi^2 = \frac{n(a_1 b_2 - a_2 b_1)^2}{n_1 n_2 n_a n_b}$$

$$= \frac{n(a_1 b_2 - a_2 b_1)^2}{(a_1 + b_1)(a_2 + b_2)(a_1 + a_2)(b_1 + b_2)}$$

With Yates' correction

$$\chi^2_{corrected} = \frac{n\left(\left| a_1 b_2 - a_2 b_1 \right| - \frac{n}{2}\right)^2}{(a_1 + b_1)(a_2 + b_2)(a_1 + a_2)(b_1 + b_2)}$$

$$= \frac{n\left(\left| a_1 b_2 - a_2 b_1 \right| - \frac{n}{2}\right)^2}{n_1 n_2 n_a n_b}$$

2 x 3 Tables

We can use the general formula

$$\chi^2 = \sum \frac{f_i^2}{e_i} - n$$

	I	II	III	Total
A	a_1	a_2	a_3	n_a
B	b_1	b_2	b_3	n_b
Total	n_1	n_2	n_3	n

to find

$$\chi^2 = \frac{n}{n_a}\left(\frac{a_1^2}{n_1} + \frac{a_2^2}{n_2} + \frac{a_3^2}{n_3}\right) + \frac{n}{n_b}\left(\frac{b_1^2}{n_1} + \frac{b_2^2}{n_2} + \frac{b_3^2}{n_3}\right) - n$$

CHAPTER 15

TIME SERIES

15.1 TIME SERIES

A time series is a set of measurements taken at specified times. Usually the measurements are taken at equal time intervals. For example the Gross National Product (GNP) is measured every year, or every quarter, or every month.

We measure the value of y (income, temperature, etc.) at times

$$t_1, t_2, \dots$$

Here, y is a function of t

$$y = y(t).$$

15.2 RELATIVES AND INDEXES

A simple way of comparing a series of observations is to use a ratio.

EXAMPLE:

Monthly sales volumes for a six-month period are known.

Month	1	2	3	4	5	6
Sales	$20,000	$16,000	$24,000	$31,000	$29,000	$37,000

Monthly Sales Volumes

We want to compare sales in each month with the sales in a specified month, say the first month. In this case month 1 is the base period. We divide each month's sales by sales in the first month.

Month	1	2	3	4	5	6
Ratio	1.00	0.80	1.20	1.55	1.45	1.85
%	100	80	120	155	145	185

Ratios and Relatives

Usually the ratios are multiplied by 100 to give percentages. The results are called relatives.

Linked relatives show each period as a percentage of the preceding period. Thus, month 1 is used as the base period for month 2, month 2 is used as the base period for month 3, etc.

Month	1	2	3	4	5	6
Sales	$20,000	$16,000	$24,000	$31,000	$29,000	$37,000
Linked Relatives		80	150	129	94	128

Linked Relatives

Linked relatives show data for each period as a percentage of the preceding period.

Indexes

Often to describe how the price of food or total production outcome changes over a period of time we construct an index (which is a single number) that reflects the changes over a period of time.

A good example is a food price.

EXAMPLE:

Over five years the price of the following items was recorded.

Prices	Year				
	1	2	3	4	5
Milk ($/gallon)	1.68	2.05	2.01	2.23	2.42
Potatoes ($/lb)	0.72	0.64	0.81	0.80	0.86
Beef ($/lb)	1.65	2.12	1.97	2.32	2.75

To construct an index we can use a simple aggregate. That is we add up all the prices and compute relatives based on their sum. Using year 1 as the base period we find:

Year	1	2	3	4	5
Total	1.68 0.72 +1.65 4.05	2.05 0.64 +2.12 4.81	4.79	5.35	6.03
Index	100	119	118	132	149

While computing the simple aggregate we assume that all elements are equally important. It is not always the case.

To remedy the situation we use a weighted aggregate index, where weights are assigned arbitrarily or based on some system of preferences.

15.3 ANALYSIS OF TIME SERIES

Time Series $y = y(t)$ is often represented by a graph.

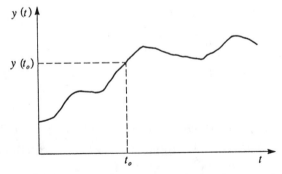

Analysis of time series shows certain characteristic measurements or variations which all of them exhibit to varying degrees.

We classify characteristic movements of time series into four categories called components of a time series.

1. Long-Term Movement

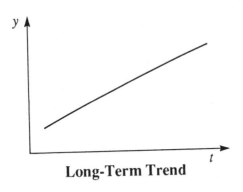

Long-Term Trend

Long-term movements refer to the general direction in which the graph of a time series is moving over a long time. For example, a long-term increase in sales volume as a result of population growth is a trend.

2. Cyclical movements or cyclical variations refer to the long-term oscillations about a trend curve. These cycles are not necessarily periodic.

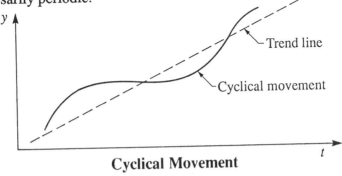

Cyclical Movement

3. Seasonal movements or seasonal variations refer to patterns, identical or almost identical, which a time series follows

during corresponding months of successive years. Month-by-month changes in sales related to holidays and changes in the weather are an example of seasonal variation.

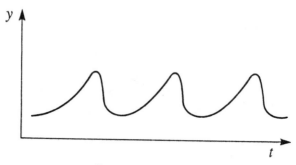

Seasonal Variation

4. Random movements refer to the motions of time series due to chance events such as war, strikes, etc. Usually it is assumed that such variations last only a short time.

In analyzing time series we assume that the variable y is a product of the variables T (trend), C (cyclical), S (seasonal), and R (random) movements.

$$y = T \times C \times S \times R$$

The analysis of time series consists of decomposition of a time series into its basic components, T, C, S, R; which are then analyzed separately.

15.4 MOVING AVERAGES

A set of numbers is given

$$y_1, y_2, y_3, \ldots$$

We define a moving average of order n to be the sequence of arithmetic means

$$\frac{y_1 + \ldots + y_n}{n}, \; \frac{y_2 + \ldots + y_{n+1}}{n}, \; \frac{y_3 + \ldots + y_{n+2}}{n}, \; \ldots$$

EXAMPLE:

In the sequence of natural numbers

$$1, 2, 3, 4, 5, 6, 7, \ldots$$

a moving average of order 3 is given by the sequence

$$\frac{1 + 2 + 3}{3} = 2, \; \frac{2 + 3 + 4}{3} = 3, \; \frac{3 + 4 + 5}{3} = 4,$$

$$\frac{4 + 5 + 6}{3} = 5, \ldots$$

which is the sequence of natural numbers starting with the number 2.

If data are supplied annually or monthly we talk about an n year moving average or n month moving average.

In general moving averages tend to reduce the amount of variation in a set of data.

In the case of time series it is called smoothing of time series. We eliminate unwanted fluctuations.

Trends

There are several methods of estimation of trend.

1. The method of least squares is used to find an appropriate trend line or trend curve.

221

2. Fitting a trend line or trend curve by looking at the graph is called the freehand method.

3. We can eliminate the cyclical, seasonal and irregular patterns by using moving averages of the proper orders. What is left is the trend movement. This method is called the moving average method.

4. We can separate the data into two parts, possibly equal, and average the data in each part. The two points obtained define a trend line. This method is called the method of semi-averages.

REA's **Problem Solvers**

The "PROBLEM SOLVERS" are comprehensive supplemental textbooks designed to save time in finding solutions to problems. Each "PROBLEM SOLVER" is the first of its kind ever produced in its field. It is the product of a massive effort to illustrate almost any imaginable problem in exceptional depth, detail, and clarity. Each problem is worked out in detail with a step-by-step solution, and the problems are arranged in order of complexity from elementary to advanced. Each book is fully indexed for locating problems rapidly.

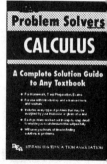

ADVANCED CALCULUS
ALGEBRA & TRIGONOMETRY
AUTOMATIC CONTROL
   SYSTEMS/ROBOTICS
BIOLOGY
BUSINESS, ACCOUNTING, & FINANCE
CALCULUS
CHEMISTRY
COMPLEX VARIABLES
COMPUTER SCIENCE
DIFFERENTIAL EQUATIONS
ECONOMICS
ELECTRICAL MACHINES
ELECTRIC CIRCUITS
ELECTROMAGNETICS
ELECTRONIC COMMUNICATIONS
ELECTRONICS
FINITE & DISCRETE MATH
FLUID MECHANICS/DYNAMICS
GENETICS
GEOMETRY

HEAT TRANSFER
LINEAR ALGEBRA
MACHINE DESIGN
MATHEMATICS for ENGINEERS
MECHANICS
NUMERICAL ANALYSIS
OPERATIONS RESEARCH
OPTICS
ORGANIC CHEMISTRY
PHYSICAL CHEMISTRY
PHYSICS
PRE-CALCULUS
PSYCHOLOGY
STATISTICS
STRENGTH OF MATERIALS &
   MECHANICS OF SOLIDS
TECHNICAL DESIGN GRAPHICS
THERMODYNAMICS
TOPOLOGY
TRANSPORT PHENOMENA
VECTOR ANALYSIS

*If you would like more information about any of these books,
complete the coupon below and return it to us or visit your local bookstore.*

RESEARCH & EDUCATION ASSOCIATION
61 Ethel Road W. • Piscataway, New Jersey 08854
Phone: (908) 819-8880

Please send me more information about your Problem Solver Books

Name _____

Address _____

City _____ State _____ Zip _____